T0135623

# Fouling of low-pressure membranes by municipal wastewater: identification of principal foulants and underlying fouling mechanisms

vorgelegt von
Diplom-Ingenieurin
CLAUDIA LAABS

Von der Fakultät III – Prozesswissenschaften
der Technischen Universität Berlin
zur Erlangung des akademischen Grades

**Doktorin der Ingenieurwissenschaften**
**– Dr.-Ing. –**

genehmigte Dissertation

Promotionsausschuss:
Vorsitzender: Prof. Dr.-Ing. Matthias Kraume (Technische Universität Berlin)
Berichter: Prof. Dr.-Ing. Martin Jekel (Technische Universität Berlin)
Berichter: Prof. Dr. Gary Amy (University of Colorado)

Tag der wissenschaftlichen Aussprache: 28. Mai 2004

Berlin 2004

D 83

Bibliografische Information Der Deutschen Bibliothek

Die Deutsche Bibliothek verzeichnet diese Publikation in der Deutschen
Nationalbibliografie; detaillierte bibliografische Daten sind im Internet über
http://dnb.ddb.de abrufbar.

ISBN 3-8325-0643-8

Logos Verlag Berlin
Comeniushof, Gubener Str. 47,
10243 Berlin
Tel.: +49 030 42 85 10 90
Fax: +49 030 42 85 10 92
INTERNET: http://www.logos-verlag.de

# Acknowledgements

I would like to thank my two advisers Prof. G. Amy and Prof. M. Jekel for the opportunity to conduct my research under their guidance. I really appreciated to work in both research groups, at the University of Colorado in Boulder (USA) and at the Technical University Berlin (Germany). I thank Prof. M. Kraume for being the chairman of my examination committee.

I would like to thank Sandra Rosenberger and Boris Lesjean from the KompetenzZentrum Wasser Berlin, and Regina Gnirß from the Berliner Wasser Betriebe for the good cooperation within the MBR study. For their help in the effluent organic matter isolation, I thank George Aiken, Jerry Leenheer, and Jamie Weishaar from the United States Geological Survey.

Anjou Recherche is kindly acknowledged for the financial support within different projects on membrane fouling.

Without the kindness and help of the staff at the City of Boulder wastewater treatment plant and at the wastewater treatment plant Ruhleben this research would not have been possible as they supplied me with cubic meters of wastewater. I would also like to thank Gisela Sosna for performing stirred cell experiments and other analyses, and Angelika Kerstin for the shared time in front of the LC-OCD system.

I am grateful for the time spent together with my colleagues in Boulder and in Berlin, for their help and friendship. I would like to specially thank Dave Foss, Hélène Habarou, Chalor Jarusutthirak, NoHwa Lee and Jochen Schumacher.

Last but not least, I would like to express my love and gratefulness to my husband Holger for his love, patience, and support. He always brought back a smile on my face when nothing seemed to work.

# Table of contents

# Abbreviations

| | |
|---|---|
| A | membrane constant, i.e. hydraulic permeability [m³/m²sbar] |
| $A_m$ | membrane area [m²] |
| AOX | adsorbable organic halogens |
| ATR-FTIR | Attenuated Total Reflectance – Fourier Transform Infra-Red spectroscopy |
| Bld | Boulder |
| Bln | Berlin |
| BOD | biochemical oxygen demand |
| BSA | bovine serum albumin |
| c | centi |
| c | concentration [mg/L] |
| C | carbon |
| CA/CN | membrane material made from a mixture of cellulose acetate and cellulose nitrate |
| CAS | conventional activated sludge |
| CDOC | chromatographable dissolved organic carbon |
| Cl⁻ | chloride |
| COD | chemical oxygen demand |
| CU | University of Colorado at Boulder, USA |
| d | days |
| D | Dalton |
| DOC | dissolved organic carbon |
| DOM | dissolved organic matter |
| EDTA | ethylenediaminetetraacetate |
| EfOM | effluent organic matter |
| EPS | extracellular polymeric substances |
| f | fractional amount of total protein that contributes to deposit growth |
| g | gram |
| h | hour |
| HCl | hydrochloric acid |
| HOC | hydrophobic organic carbon |
| HPLC | high pressure liquid chromatography |
| HRT | hydraulic retention time [h] |
| HS | humic substances |
| IC | inorganic carbon |
| IMS | integrated membrane systems |
| IX | ion-exchanger |
| J | permeate flux [L/hm²] |
| k | kilo |
| k | rate constant |

| | |
|---|---|
| L | litre |
| L | capillary length [m] |
| LC-OCD | liquid chromatography-organic carbon detection |
| m | milli |
| m | meter |
| M | molar [mol/L] |
| MBR | membrane bio-reactor |
| MF | microfiltration |
| MFI | modified fouling index |
| MLSS | mixed liquor suspended solids [g/L] |
| MW | molecular weight [g/mol] |
| MWCO | molecular weight cut-off [D] |
| n | nano |
| n | filtration constant |
| N | normal |
| NaOH | sodium hydroxide |
| NDIR | non-dispersive infra-red detector |
| NF | nanofiltration |
| $NH_4^+$-N | ammonia-nitrogen |
| NMR | nuclear magnetic resonance |
| $NO_3^-$-N | nitrate-nitrogen |
| NOM | natural organic matter |
| OC | organic carbon |
| p | pressure [bar] |
| PAN | polyacrylonitrile |
| PCA | principal component analysis |
| PEG | polyethylene glycol |
| POC | particulate organic carbon |
| POM | particulate organic matter |
| PP | pilot plant |
| PS | polysaccharides |
| PVDF | polyvinylidene fluoride |
| PWP | pure water permeability [L/m²hbar] |
| Q | flow rate [m³/s] |
| R | resistance [1/m] |
| R' | specific protein layer resistance [1/m] |
| RC | regenerated cellulose |
| RO | reverse osmosis |
| s | seconds |
| S | Siemens |
| S | volume specific surface [m²/m³] |
| SBR | sequencing batch reactor |
| SDI | silt density index |

| sec | seconds |
| --- | --- |
| SEC | size exclusion chromatography |
| SMP | soluble microbial products |
| SRT | solids or sludge retention time [d] |
| SUVA | specific UV absorption at a wavelength of 254 nm = $UV_{254}$/DOC [L/mg m] |
| t | filtration time [h] |
| TKN | total Kjeldahl nitrogen |
| TMP | transmembrane pressure [bar] |
| TOC | total organic carbon [mg C/L] |
| TP | total phosphorous |
| TU | Technical University Berlin, Germany |
| UF | ultrafiltration |
| UV | ultraviolet |
| $UV_{254}$ | ultraviolet absorption at a wavelength of 254 nm |
| V | volume [L] |
| WWTP | wastewater treatment plant |

**Indices**

| 0 | initial |
| --- | --- |
| a | adsorptive fouling |
| b | bulk |
| c | cake layer |
| c | concentrate |
| cp | concentration polarisation |
| f | feed |
| m | membrane |
| p | permeate |
| pr | protein |
| v | volumetric |

**Greek letters**

| $\alpha$ | pore blockage parameter |
| --- | --- |
| $\Delta$ | difference, e.g. $\Delta p$ = pressure difference |
| $\epsilon$ | porosity |
| $\eta$ | dynamic viscosity [kg/ms] |
| $\mu$ | micro |
| $\mu$ | viscosity of water |
| $\Pi$ | osmotic pressure |

**Glossary**

| colloids | substances retained in a dialysis bag in the isolation protocol |
| --- | --- |
| concentrate | sample remaining in stirred cell at the end of the filtration test |
| feed | water sample used for stirred cell experiments |

| | |
|---|---|
| filtrate | aqueous sample from the membrane reactor filtered over paper filter to separate the activated sludge from the water phase |
| HPO-A | hydrophobic acids; substances adsorbing onto XAD-8 and eluting with NaOH |
| HS peak | humic substances peak including humic substances and humic hydrolysates |
| lyophilization | freeze-drying (in this case: of isolated effluent organic material) |
| perm | permeate collected during stirred cell experiments |
| permeate | aqueous sample taken after the membrane of the MBR pilot plants |
| PP 1 | pilot plant 1 operated in a pre-denitrification configuration |
| PP 2 | pilot plant 2 operated in a post-denitrification configuration without additional dosing of an external carbon source |
| PS peak | so-called polysaccharides peak including organic colloids, polysaccharides, and proteins |
| retentate | sample remaining in stirred cell at the end of the filtration test |
| TPI-A | transphilic acids; substances adsorbing onto XAD-4 and eluting with NaOH |

# Zusammenfassung

Im Bereich der kommunalen Abwasserbehandlung finden Mikrofiltrationsmembranen (MF) und Ultrafiltrationsmembranen (UF) zunehmend Anwendung. Zum einen werden sie zur weitergehenden Reinigung des Kläranlagenablaufs eingesetzt. Zum anderen kommen sie in Form von Membranbelebungsreaktoren (MBR) zum Einsatz. Die Hauptproblematik bei der Anwendung von Niedrigdruckmembranen (MF und UF) zur kommunalen Abwasserreinigung liegt nach wie vor im Fouling (Verblocken) der Membranen durch organisches Material. Das Fouling führt dabei zu einer Verminderung der Effizienz der Membranstufe durch die Verringerung des transmembranen Flusses oder durch den Anstieg der transmembranen Druckdifferenz über die Zeit.

Thema dieser Arbeit ist die Untersuchung des Foulingpotenzials von kommunalem Abwasser für Niedrigdruckmembranen. Das Ziel ist die Identifizierung und Charakterisierung der Fouling verursachenden Substanzen sowie der zu Grunde liegenden Foulingmechansimen. Das Foulingverhalten unterschiedlicher Wässer (Kläranlagenablauf, verschiedene Fraktionen des Kläranlagenablaufs, MBR Zulauf und Permeat) wurde für fünf MF und UF Membranen in Rührzellenversuchen untersucht. Zusätzlich wurden zwei MBR Pilotanlagen über ein Jahr hinweg beprobt, um mögliche Zusammenhänge zwischen Beschaffenheit des Abwassers, Betriebsparametern und Membranfouling zu erkennen.

Die Rührzellenversuche zeigten, dass lediglich 10 % der im Kläranlagenablauf vorhandenen organischen Substanzen zum Fouling der Membranen beitragen. Bei den Fouling verursachenden Wasserinhaltsstoffen handelt es sich um hydrophile organische Substanzen, die bei einer Größenausschlußchromatographie im sogenannten PS Peak eluieren. Dieser Peak umfasst neben Polysacchariden auch organische Kolloide und Proteine.

Die Beprobung der Membranbelebungsreaktoren ergab, dass hier ebenfalls Substanzen, die im PS Peak eluieren, für das Fouling ausschlaggebend sind. Huminstoffe und organische Säuren sind nur von untergeordneter Bedeutung. Die Konzentration des organischen Kohlenstoffs im PS Peak nahm mit zunehmender Temperatur des Abwasser ab. Ein Einfluss des Schlammalters (die beiden Pilotanlagen wurden jeweils mit Schlammaltern von 8 Tagen und 15 Tagen gefahren) auf den PS Peak konnte dagegen nicht festgestellt werden. Vielmehr sind Stresssituationen (z. B. Sauerstoffmangel) von entscheidendem Einfluss auf den PS Peak und damit das Fouling, da sie eine vermehrte Produktion extrazellulärer polymerer Substanzen (EPS) durch die Mikroorganismen des belebten Schlammes bewirken.

Sequenzielle Filtrationsversuche ergaben, dass die im PS Peak eluierenden Substanzen etwa 10-100 nm groß sind. Dies entspricht dem Größenbereich von Kolloiden (1 nm – 1 µm). Die der Membranfiltration dieser kolloidalen Substanzen zu Grunde liegenden Foulingmechanismen unterscheiden sich je nach Membrantyp: Bei UF Membranen kommt es zur Bildung eines Filterkuchens, während MF Membranen durch eine Kombination aus Porenverengung und Bildung eines Filterkuchens verblocken.

# Abstract

Microfiltration (MF) and ultrafiltration (UF) membranes have a wide range of applications in the wastewater industry. On one hand, they are used to polish wastewater treatment plant (WWTP) effluents. On the other hand, they are directly integrated in the biological treatment process in the form of membrane bio-reactors (MBR). The major challenge of low-pressure membrane filtration remains the fouling of the membranes by organic matter which prevents an efficient operation by reducing the overall transmembrane flux and/or increasing the transmembrane pressure over time. This research investigated the fouling potential of municipal wastewater in low-pressure membrane filtration. The aim was to identify the principal foulants and underlying fouling mechanisms. Bench-scale stirred cell experiments were conducted to assess the fouling behaviour of various feed waters (bulk effluents, effluent isolates, MBR feed and permeate) tested with five MF and UF membranes. Additionally, two MBR pilot plants have been monitored over the course of one year in order to detect possible correlations between the feed water characteristics, operational parameters, and membrane fouling.

Although stirred cell experiments showed significant flux decline, only about 10 % of the effluent organic matter (EfOM) present in wastewater treatment plant effluents are responsible for the observed fouling of the low-pressure membranes (MF and UF). The principal foulants can be attributed to hydrophilic organic matter eluting in the so-called polysaccharides (PS) peak during size exclusion chromatography with on-line DOC detection; this PS peak contains organic colloids, polysaccharides, and larger proteins. This is supported by stirred cell tests with isolated colloids.

As for WWTP effluent membrane filtration, the PS peak negatively impacts the membrane performance of membrane bio-reactors. In contrast, humic substances and organic acids play only a minor role in the fouling. The organic carbon concentration of the PS peak was influenced by the wastewater temperature; with increasing temperature the PS peak decreased. The MBR pilot plants were operated at two different sludge retention times (SRT) of eight and fifteen days. However, no correlation could be found between the sludge retention time (SRT) and the magnitude of the PS peak. Additionally, stress situations (e.g. oxygen deprivation) for the biomass appeared to impact the PS peak as they influence the amount of extracellular polymeric substances (EPS) produced by the microorganisms.

Sequential filtration stirred cell experiments revealed that the compounds eluting in the PS peak have sizes of tens to hundreds of nanometers which is the typical size range of colloids (definition of colloids: 1 nm – 1 μm). These colloidal substances foul low-pressure membranes either by forming a cake layer on top of the membrane surface (as is the case for ultrafiltration membranes) or by a combination of pore constriction and cake or gel layer formation (as observed for microfiltration membranes).

# Chapter 1     Introduction

## *1.1  Introduction*

Facing the growing worldwide demand for safe drinking water and, at the same time, the diminishing amount of suitable fresh water resources, it becomes more and more crucial to ensure a sustainable development and use of the fresh water sources. The implementation of local or regional wastewater reuse systems allows to close the gap in the water cycle between wastewater and drinking water and thereby, contributes to the sustainable use of the available fresh water resources. In industrialized countries, most of the domestic wastewater is treated in centralised public wastewater treatment plants (WWTP) by mechanical, biological and chemical processes. The first step (mechanical clarification) reduces the amount of organic material in the form of particulates by 30 to 35 % [Schwedt 1996] while most of the dissolved organic substances are eliminated in the secondary treatment by biological processes, i.e. activated sludge or trickling filter. A third process train including biological and/or chemical processes reduces nutrients such as nitrogen and phosphorus. In the wake of ever more stringent regulations with respect to the effluent quality of wastewater treatment plants, membrane filtration is becoming an attractive treatment alternative to conventional treatment for ensuring high quality effluents that can be used in wastewater reclamation and reuse systems.

Membranes are able to remove dissolved compounds, colloids, and particles from the feed stream to reduce the load of bulk and trace organic material as well as inorganic contaminants depending on the pore size or molecular weight cut-off (MWCO) of the membrane material. Although the first ultrafiltration membranes were developed by the 1930s, reverse osmosis membranes were the first to be used in the water industry for water production from seawater (desalination) starting in the 1960s [Anselme & Jacobs 1996]. Low-pressure membranes, i.e. microfiltration and ultrafiltration membranes, made their entrance in the water and wastewater industry in the 1980s. An increase in installed membrane capacity in the water and wastewater industry over the last decade can be observed [EAWAG 2000]. As the installed membrane capacity rises, the production costs fall and new knowledge and experience with membrane filtration of water and wastewater is gained. This induces further installations of membrane operations. Over the last ten to fifteen years microfiltration (MF) and ultrafiltration (UF) membranes have become a technically and economically feasible alternative for water and wastewater treatment.

Low-pressure membrane filtration can be integrated at different points in the process of municipal wastewater treatment. Microfiltration and ultrafiltration membranes are, for example, placed into the biological treatment. Especially in small decentralised wastewater treatment plants (WWTP) this membrane bio-reactor technology (MBR) has advantages over a conventional treatment process, i.e. better effluent quality and at the same time smaller footprints. In contrast, larger WWTP prefer to add low-pressure membrane filtration as a polishing step after the secondary clarifiers. This allows them to meet more stringent regulations with respect to the effluent quality without the need to change the whole treatment process. Furthermore, low-pressure membranes are used in integrated membrane systems (IMS) in order to purify secondary/tertiary effluents for indirect or direct potable reuse. A well-known indirect potable water reuse system is the Water Factory 21 facility in Orange County (California, USA) with its reverse osmosis operating since 1976. However, Water Factory 21 will soon be replaced with the groundwater replenishment (GWR) system. The GWR uses an integrated membrane system including microfiltration membranes followed by reverse osmosis. Secondary effluent from a wastewater treatment plant is treated by the IMS prior to UV oxidation and injection into the groundwater aquifer. The GWR project reduces the amount of secondary effluent being discharged into the Pacific Ocean while at the same time recharging the groundwater and thereby replenishing the salt water intrusion barrier [Everest et al. 2002].

The major challenge of membrane filtration remains the fouling of membranes which prevents an efficient operation. Four types of fouling can be distinguished:
  - scaling due to inorganic compounds (salts),
  - biofouling due to the attachment and growth of microorganisms,
  - organic fouling due to dissolved and colloidal organic matter, and
  - particulate fouling due to organic and/or inorganic particles.
However, scaling is only of relevance in reverse osmosis and nanofiltration plants. In low-pressure membrane operations, divalent ions, responsible for scale formation, permeate the membranes. More crucial is the organic fouling of microfiltration and ultrafiltration membranes.

Besides the natural organic matter (NOM) causing fouling in drinking water treatment plants, membrane operations of municipal wastewater have to deal with additional organic substances such as disinfection by-products (DBP) from the drinking water treatment, synthetic organic compounds (SOC) added by the consumer, and organic substances produced by the microorganisms during biological wastewater treatment [Drewes and Fox 1999]. Depending on the membrane application, organic substances produced by microorganisms are referred to as soluble microbial products (SMP, wastewater effluent membrane filtration) or extracellular polymeric substances (EPS, MBR applications). However, Laspidou and Rittmann [2002] indicated that SMP and soluble EPS are referring to the same group of compounds.

Thus far, strategies with regards to fouling control and membrane cleaning are primarily based on trial-and-error. A more in-depth knowledge is needed about the substances causing organic fouling in low-pressure membrane filtration as well as on the relevant fouling mechanisms. This would enable a better choice concerning membrane selection as well as cleaning strategies, thereby reducing chemicals and extending membrane life.

Hence, the topic of this research is the fouling potential of municipal wastewater organic matter in low-pressure membrane filtration (MF and UF). The aim is to identify the principal foulants in low-pressure membrane filtration of wastewater treatment plant effluents as well as in membrane bio-reactors integrated in wastewater treatment plants. Furthermore, the underlying fouling mechanisms are investigated.

## 1.2    Structure of thesis

An introduction into low-pressure membrane filtration and its basic concepts is given in Chapter 2 followed by an overview of the literature on fouling of low-pressure membranes by municipal wastewater organic matter.

A description of the stirred cell testing equipment and the membrane bio-reactor pilot plants, the characterisation of the feed waters, and an explanation of analytical measurements and procedures (isolation scheme, size exclusion chromatography, Fourier transform infra-red spectroscopy, organic carbon detection, $UV_{254}$ absorption) is given in Chapter 3.

The results of this research are presented and discussed in Chapters 4-7 starting with the fouling potential of bulk wastewater treatment plant effluents in Chapter 4. An analysis of the obtained effluent isolates and their fouling properties and character is carried out in Chapter 5. Chapter 6 identifies foulants in membrane bio-reactors and assesses their behaviour in two pilot plants treating municipal wastewater with nitrification/ denitrification and biological phosphorous removal. Flux data derived from stirred cell experiments as well as the operational data from the two pilot plants are analysed using principal component analysis, a statistical analysis method to determine the parameters influencing most significantly fouling of low-pressure membranes (Chapter 7.1). Furthermore, the flux data of the stirred cell experiments is examined with mathematical models in order to describe the underlying fouling mechanisms (Chapter 7.2). A concluding discussion of the thesis' main findings and some recommendations for future research can be found in Chapter 8.

# Chapter 2    Theory    and    background    of    low-pressure membranes

## 2.1    Pressure-driven membrane filtration

The basic principle of all membrane operations is the separation of a mixture of substances with a selective thin film. The separation is based on the differing transport resistance of the single components. The transport of matter through the selective barrier is caused by a chemical potential difference between the two phases, i.e. feed and permeate. The driving force of the chemical potential difference can be either a difference in:

- activity as is the case for dialysis, pervaporation, membrane stripping
- electrical potential as is the case for electrodialysis, or
- pressure as is the case in microfiltration, ultrafiltration, nanofiltration, reverse osmosis, pervaporation, and membrane stripping [Aptel and Buckley 1996].

The different membrane operations are used to concentrate, fractionate or purify solutions. In the wastewater industry, the solution of interest is raw wastewater that has been pretreated to varying degrees and needs further purification. In general, pressure driven membranes are employed for this purification step, i.e. microfiltration, ultrafiltration, nanofiltration or reverse osmosis membranes. In this research microfiltration and ultrafiltration membranes have been investigated.

### 2.1.1    Membrane classification and characterisation

#### Membrane classification

The membrane filtration separates the feed stream into a permeate stream and a concentrate or retentate stream. The permeate consists of the purified water and any substances that are able to pass the membrane barrier. All substances that are rejected by the membrane are found in the concentrate (Figure 2.1). The membranes work as selective barrier to specific compounds present in wastewater, e.g. organic colloids, bacteria, viruses, humic substances, ions. Depending on the type of membrane used and on the applied driving force, different kinds of compounds can be retained by membranes.

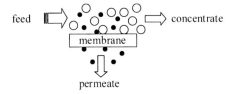

feed                                        concentrate

permeate

**Figure 2.1**    Membrane separation of wastewater (feed) into a permeate and a
                  concentrate stream

In pressure-driven membrane filtration the driving force is an applied pressure difference
across the membrane. Typically, four membrane types are distinguished: reverse osmosis
(RO), nanofiltration (NF), ultrafiltration (UF), and microfiltration (MF). A classification of
the membranes can be done according to their separation range (molecular weight cut-off
or pore size) and the applied transmembrane pressure (Figure 2.2):
*Reverse osmosis* membranes retain monovalent ions and low molecular weight solutes. The
rejection of the salts present in the water/wastewater leads to a high osmotic pressure
which has to be overcome by the applied pressure. Generally, at least twice the osmotic
pressure is needed for sufficient flow rates [Aptel and Buckley 1996]. The applied pressure
for these dense membranes ranges from 10-100 bar. Their primary application is sea water
desalination.
*Nanofiltration* membranes lie between reverse osmosis and ultrafiltration membranes with
regards to the applied pressure (5-30 bar) as well as their pore size. The pore sizes are in
the nanometer range placing the membranes on the border of porous and dense
membranes. The necessary pressure is lower than in RO operations because monovalent
ions pass the membrane, thereby reducing the osmotic pressure. Additionally, the
membrane surface of nanofiltration membranes is negatively charged. This results in a
good rejection of divalent cations (calcium and magnesium) and therefore, they are used
for water softening.
*Ultrafiltration* membranes have a porous structure and reject macromolecules, colloids,
and particulates. This includes the retention of microorganisms, bacteria, and viruses which
means that these membranes can be used for disinfection purposes [Aptel and Buckley
1996]. Due to the fact that most inorganic ions pass the membrane, the applied pressure (1-
10 bar) is directly used for the filtration of the solution without the need to overcome any
osmotic backpressure.
*Microfiltration* membranes are the loosest membranes with pore sizes of 0.05 µm to 5 µm.
They are mainly used for particle and microbial removal, although viruses are smaller than
the pore size and can be able to pass the membrane (see Figure 2.2). Nevertheless, the
formation of a cake layer on the membrane surface oftentimes leads to an additional
removal capacity for smaller components  [Jacangelo and Buckley 1996]. The operating
pressure lies between 0.1-3 bar.

In the wastewater industry all four types of membranes are used in various applications. Microfiltration and ultrafiltration membranes are, for example, used for microbial removal and as a polishing step in wastewater treatment plant effluent filtration [Roorda 2004]. Integrated membrane systems (IMS) combine a high-pressure membrane (RO, NF) with a low-pressure membrane (MF, UF) placed upstream to protect the high-pressure membranes from foulants. These combinations are often found in wastewater reclamation and reuse applications such as the groundwater replenishment system in Orange County (formerly Water Factory 21) [Everest et al. 2002]. In membrane bio-reactors (MBRs), low-pressure membranes replace the secondary clarifier in wastewater treatment plants retaining the biomass in the biological reactors and thereby, minimizing the loss of biomass and ensuring a high quality effluent free of microorganisms and turbidity [Manem and Sanderson 1996].

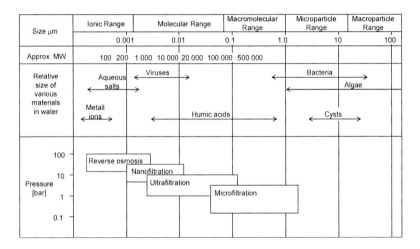

**Figure 2.2**    Types of membranes, applied pressure ranges, and removal capacities, adapted from [Anselme and Jacobs 1996] and [Voßenkaul and Rautenbach 1997]

*Membrane characterisation*

Membranes can be characterised according to their retention capabilities and to material composition. The retention capability is supplied by the manufacturers in terms of molecular weight cut-off (MWCO) for RO, NF, and UF. The molecular weight cut-off is defined as the molar mass of a solute that is 90 % retained by the membrane. Solutes for the assessment of the MWCO are polyethylene glycol (PEG), dextran, and globular proteins. However, no international standard exists for the determination of the MWCO and therefore, it is difficult to compare membranes from different manufacturers based on the specified MWCO alone [Anselme and Jacobs 1996]. Microfiltration membranes, on the other hand, are characterised by their pore size.

Regardless of a characterisation by MWCO or pore size, it should be noted that the mechanisms involved in rejection are very complex. Besides the membrane, the feed water characteristics as well as the operating conditions influence the rejection and can themselves change over time due to fouling and membrane aging [Anselme and Jacobs 1996]. Table 2.1 gives an overview of possible parameters influencing rejection.

**Table 2.1**    Physical and chemical parameters influencing membrane rejection, adapted from [Anselme and Jacobs 1996]

| Membrane-related factors | Feed water characteristics | Operating conditions |
| --- | --- | --- |
| MWCO/ pore size | pH | cross-flow vs. dead-end |
| pore size distribution | ionic strength | cross-flow velocity |
| surface charge | mineral content | transmembrane pressure |
| roughness | salts/ metallic oxide precipitation | laminar/ turbulent flow |
| material adsorption | organic content (organic fouling) | |
| capacity | particle content (cake formation) | |

The membrane material can be of organic (polymers) or inorganic (metals, ceramics, glasses) nature. Due to lower manufacturing costs, most membranes are made of synthetic polymers. Typical membrane materials are regenerated cellulose, cellulose acetate, polyvinylidene fluoride, polyamide, polyacrylonitrile, polysulfone, polyethersulfone, polypropylene, etc. (see Aptel and Buckley [1996] for molecular structures of the corresponding monomers).

Oftentimes, the more hydrophobic membrane materials are modified during the manufacturing process to enhance the hydrophilicity of the membrane surface. Hence, some membrane suppliers state the contact angle as measure for the hydrophobicity of the membrane surface. A low contact angle represents a hydrophilic material while a higher contact angle is typically found with hydrophobic membrane surfaces, e.g. > 50°.

### 2.1.2  Membrane modules
Membranes come in two different geometries: as cylindrical membranes and as flat sheet membranes. Both types are available in various modules. A module is defined as single operational unit into which the membranes are engineered for use [Anselme and Jacobs 1996]. Flat sheet membranes are either used in plate and frame, cushion or spiral wound modules. In *plate and frame* modules the flat sheet membranes are stacked with support plates. They are easy to disassemble for manual cleaning and replacement of the individual membrane sheet. A disadvantage is their low packing density and high pressure drop [Aptel and Buckley 1996]. *Spiral wound* modules have higher packing densities and a

lower head loss. However, they are more difficult to clean and require a higher amount of feed water pre-treatment. They are manufactured by connecting one or more membrane envelopes to a perforated tube. An envelope consists of two flat sheet membranes that enclose a porous permeate collector sheet. The envelopes are separated from each other by the feed spacer and rolled up around the perforated tube. The permeate is collected in the perforated tube and extracted at either end of the module [Aptel and Buckley 1996]. *Cushion* modules are similar to spiral wound modules in that envelopes are formed from two membrane sheets. However, the envelopes are not rolled up but sealed on all four sides and the permeate is drained through a hole that is sealed against the feed stream. The main disadvantage of cushion modules is that the membrane must be able to withstand welding and glueing/sealing. This holds true for spiral wound modules, too. The advantages are few seals and low fouling [Rautenbach 1997].

Cylindrical membranes can be engineered as tubular modules or hollow-fiber modules. The *hollow-fiber* modules are further differentiated into hollow fine fibers for RO applications and hollow capillary fibers used in MF and UF applications. The difference between these two is the inner diameter which is larger for capillary fibers (350-1000 µm) [Anselme and Jacobs 1996] than for fine fibers. In hollow-fiber modules the membranes are self-supporting due to the small inner diameter. This allows for the possibility of backflushing these modules which is an advantage when it comes to fouling control. Furthermore, the self-supporting structure results in the possibility of having either an inside-out flow pattern (feed is inside the fibers) or an outside-in flow (feed is outside the fibers while the permeate is collected inside). Another advantage is the high packing density with thousands to millions of fibers bundled together [Aptel and Buckley 1996].

With the *tubular* modules, the cylindrical membrane is inside a support tube which has an internal diameter of 6-40 mm. In the case of inorganic membranes, multi-channel ceramic support tubes can be used with up to 19 flow channels. Nevertheless, this results in a very low packing density and high capital cost. The advantage is an easy to clean module with defined hydrodynamics [Aptel and Buckley 1996].

**Table 2.2**   Membrane modules used for the four pressure-driven membrane applications: MF, UF, NF, RO according to [Rautenbach 1997]

|                        | MF | UF | NF | RO |
|------------------------|----|----|----|----|
| Plate and frame        | x  | x  |    | x  |
| Cushion                |    | x  | x  | x  |
| Spiral wound           |    | x  | x  | x  |
| Tubular                | x  | x  |    | x  |
| Hollow fine fiber      |    |    |    | x  |
| Hollow capillary fiber | x  | x  |    |    |

Table 2.2 gives an overview of the types of modules typically used for the four pressure-driven membrane operations. Most often spiral wound and hollow fine fiber modules are used for high-pressure membranes. In low-pressure membrane filtration, hollow capillary fiber modules are the modules of choice.

Membranes can be operated in the dead-end mode or the cross-flow mode. In the dead-end mode, there is no retentate stream. Thus, any compounds present in the feed water either pass the membrane or accumulate on the membrane surface. In the cross-flow mode, the feed stream is directed over the membrane in a tangential flow pattern. This ensures that most of the rejected substances are carried away from the membrane surface with the concentrate stream. In most cases, the cross-flow mode is employed in order to minimize concentration polarization and fouling.

In this study, MF and UF flatsheet membranes are used for stirred cell experiments and two MF hollow-fiber modules are operated in the membrane bio-reactor pilot plants. Stirred cell experiments are carried out in the dead-end mode, although, a magnetic stirrer is used to minimize concentration polarization and to mimic cross-flow regimes. The cross-flow mode is used in the two pilot plants.

### 2.1.3  Permeate flux and solute transport

Microfiltration and ultrafiltration membranes are pressure-driven membrane operations. Thus, the driving force is a pressure difference $\Delta p$ or transmembrane pressure difference (TMP) between the feed ($p_{feed}$) and permeate ($p_{permeate}$) side:

$$\Delta p = p_{feed} - p_{permeate} \qquad \text{(eq. 2.1)}$$

Equation 2.1 holds true for dead-end or direct filtration membrane operations. For membrane modules that are operated in the cross-flow mode only an average TMP can be calculated due to the pressure drop across the module. In this case, equation 2.2 can be used [Jacangelo and Buckley 1996]:

$$\Delta p = \frac{p_{feed} - p_{concentrate}}{2} - p_{permeate} \qquad \text{(eq. 2.2)}$$

For the efficient application of membrane filtration the permeate flux J is of most importance. It is defined as the permeate flow rate $Q_p$ (= permeate volume $V_p$ per filtration time t) divided by the membrane area $A_m$:

$$J = \frac{Q_p}{A_m} = \frac{V_p}{A_m t} \qquad \text{(eq. 2.3)}$$

In order to assess the solute transport through a membrane several models exist. In the case of reverse osmosis (dense) membranes the solution-diffusion model is commonly used. The solution transport in dense membranes takes place according to the diffusion laws (Fick's law). Hence, the solute must be able to diffuse into the membrane structure in order to be transported through it. This leads to a concentration gradient of the solute across the membrane while the solvent (water) flux is a function of the transmembrane pressure (Figure 2.3 right). The general form for the transport equation is:

$$\text{flux} = \text{concentration} * \text{mobility} * \text{force} \qquad \text{(eq. 2.4)}$$

More detailed information on and a mathematical description of the solution-diffusion model can be found in Rautenbach [1997], and Wijmans and Baker [1995]. Nanofiltration membranes take an intermediate position between solution-diffusion and pore-flow membranes. Additionally, their surface charge plays an important role in the solute rejection characteristics. Special models exist for these membranes, e.g. the extended Nernst-Planck equation [Tsuru et al. 1991].

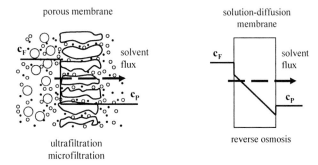

**Figure 2.3**    Concentration distribution of a solute according to the pore-flow model (left) and the solution-diffusion model (right); idealized, i.e. without concentration polarization, adapted from [Rautenbach 1997]

Low-pressure membranes (MF and UF), on the other hand, are porous membranes with a convective transport through the membrane pores. The adequate model for solution transport is the pore-flow model. In this model the membrane is defined as a system of parallel capillaries. It is assumed that the solvent (water) flow through these capillaries can be described by the Hagen-Poiseuille law. Hence, a linear relationship exists between permeate flux and applied pressure difference which is given by the membrane constant A (equation 2.5). The membrane constant can be calculated using the Carman-Kozeny relationship from filtration theory [Rautenbach 1997].

$$J = A * \Delta p \qquad A = \frac{\varepsilon^3}{\eta(1-\varepsilon)^2 S^2 2L} \qquad \text{(eq. 2.5)}$$

with

| | | |
|---|---|---|
| J | = | permeate flux [m³/m²s] |
| A | = | membrane constant or hydraulic permeability [m³/m²sbar] |
| $\Delta p$ | = | transmembrane pressure difference [bar] |
| $\eta$ | = | dynamic viscosity [kg/ms] |
| $\varepsilon$ | = | porosity |
| S | = | volume specific surface [m²/m³] |
| L | = | capillary length [m] |

The solute transport, in contrast, depends on the relationship between the pore size of the membrane and the molecular size/weight of the substances that are to be separated (Figure 2.3 left). If the compounds are of the same size as the pores, part of them will be retained and part of them will be able to pass the membrane due to the unavoidable spread in pore sizes [Rautenbach 1997].

Another way to write the relationship between pure water permeate flux J of a clean MF or UF membrane is Darcy's law:

$$J = \frac{\Delta p}{\mu \, R_m} \qquad \text{(eq. 2.6)}$$

where $\mu$ is the viscosity of water and $R_m$ is the hydraulic resistance of the clean membrane [Wiesner and Aptel 1996]. The advantage is that additional resistances due to fouling can be accounted for (resistance-in-series model):

$$J = \frac{\Delta p}{\mu \left( R_m + R_{cp} + R_c + R_a \right)} \qquad \text{(eq. 2.7)}$$

with

| | | |
|---|---|---|
| $R_m$ | = | resistance of the clean membrane |
| $R_{cp}$ | = | resistance due to concentration polarization |
| $R_c$ | = | resistance due to a cake layer |
| $R_a$ | = | resistance due to adsorptive fouling |

where each resistance term corresponds to one of the fouling mechanisms (see Section 2.1.4).

## 2.1.4  Limiting factors to the performance of membrane operations

The performance of membrane operations is a function of various parameters such as:
-   membrane material and module construction,

- characteristics of the feed water (e.g. organic carbon concentration, SUVA, temperature, viscosity),

- hydrodynamic conditions (e.g. flux, TMP, crossflow velocity),

- operational conditions (e.g. solid retention time (SRT), hydraulic retention time (HRT), oxygen concentration).

It depends primarily on the interactions between the membrane and the feed water and can be influenced to some extent by the operational and hydrodynamic conditions. Generally, concentration polarization and fouling are seen as the most important factors limiting membrane performance. Furthermore, physical aging by membrane compaction or chemical aging due to acids or bases used for cleaning can limit the membrane performance.

### *Concentration polarization*

The phenomenon of concentration polarization is inherent to membrane filtration (Figure 2.4). Due to the constant transport of feed water to the membrane surface and the selective retention of certain solutes, these solutes accumulate on and near the membrane surface. Hence, their concentration increases over the filtration time resulting in a boundary layer of higher concentration with its maximum at the membrane surface ($c_m$). The concentration build-up causes a diffusive back-transport (= $D \cdot (dc/dx)$) into the bulk feed water ($c_b$). Under steady state conditions the convective solute flow (= $J \cdot c$) is equalised by the solute flux through the membrane (= $J \cdot c_p$) plus the diffusive back-transport [Mulder 1991].

Concentration polarization is especially of concern in low-pressure membrane filtration due to the high flux that can be achieved in microfiltration and ultrafiltration. It can result in a higher fouling since the solute concentration on the membrane surface increases. At the same time the flux is decreased because of the additional resistance due to the concentration polarization layer ($R_{cp}$). Concentration polarization is controlled by the cross-flow velocity in cross-flow membrane operations or by the introduction of turbulences in dead-end membrane operations (air bubbles, stirrer).

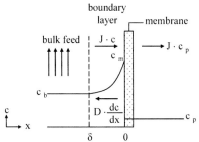

**Figure 2.4**     Concentration profile in the concentration polarization boundary layer [Mulder 1991]

The most challenging task when operating a membrane system is the fouling as it reduces the flux or increases the transmembrane pressure, respectively, depending on whether the system is operated at constant pressure or constant flux. In conjunction with the flux reduction, a shift in the effective pore size/ MWCO to smaller diameters is oftentimes observed. This can result in a MF process to display characteristics of UF membrane filtration [Rautenbach 1997]. Four types of fouling can be distinguished: scaling, biofouling, organic fouling, and particulate fouling.

### Scaling

The formation of a scale on the membrane surface can occur if dissolved salts exceed their solubility product. Typically, over-saturation is of concern in reverse osmosis and nanofiltration operations with regards to $CaCO_3$, $CaSO_4$, $BaSO_4$, $SrSO_4$, $MgCO_3$, and $SiO_2$. The formation of precipitates can be controlled over the pH range (more acidic water reduces the formation of $CaCO_3$) and/or by dosing scale inhibitors. In low-pressure membrane filtration (MF and UF), however, scaling is not of concern as the salts pass the membrane freely.

### Biofouling

Biofouling is defined as the adhesion and growth of microorganisms on the membrane surface, i.e. the formation of a biofilm, which results in a loss of membrane performance. Ridgway and Flemming [1996] state that biofilms can occur on all kinds of materials, natural and synthetic, due to the fact that bacteria have developed elaborate adhesion mechanisms. Generally, the adhesion to surfaces is achieved by extracellular polymeric substances (EPS). These are biopolymers which are produced by the microorganisms and envelop them completely thereby forming a gel-like matrix of viscous consistence. EPS consists of polysaccharides, proteins, glycoproteins, lipoproteins, and other macromolecules of microbial origin [Flemming et al. 1997].

### Organic fouling

The term organic fouling refers to a reduction in flux due to the adsorption of dissolved organic substances on the membrane surface or in its pores. The adhesion depends primarily on the chemical and electrostatic characteristics of the organic material. In drinking water membrane filtration organic fouling is due to natural organic matter (NOM) [Lee 2003]. Besides NOM low-pressure membrane filtration of municipal wastewater has to deal with synthetic organic compounds (SOC) added by the consumer and soluble microbial products (SMP) produced during the biological wastewater treatment process [Drewes et al. 1999]. The term "soluble microbial products" is basically identical with soluble EPS [Laspidou and Rittmann 2002] implying that extracellular polymeric substances can have a fouling potential by themselves outside of a biofilm. Thus, polysaccharides, proteins, and lipids may play an important role in organic fouling of membrane applications to municipal wastewater.

The term "dissolved" usually refers to any compound that passes a 0.45 μm filter. Anything that is retained by this filter is considered as being particulate. This operational definition of dissolved/particulate material does not account for colloids. Colloids are defined as organic or inorganic units with sizes between 1 nm and 1 μm and therefore, can occur in both fractions. According to Buffle and Leppard [1995a] the following compounds are colloidal in nature: $CaCO_3$, iron hydroxides, microorganisms and microbiological debris, polysaccharides, humic substances, amorphous silica, and clay. Thus, fouling by colloids might be found in all four categories: scaling, biofouling, organic fouling, and particulate fouling. The emphasis in this research is put on the investigation of organic fouling including fouling caused by organic colloids and soluble EPS.

### Particulate fouling

Small particles accumulate on the membrane surface thereby forming a filtration cake. This cake layer can increase the membrane resistance ($R_c$) adding to the overall filtration resistance. The amount of fouling caused by particulates is assessed by the silt density index (SDI) or the modified fouling index (MFI) using a 0.45 μm Millipore filter. Both indices are, for example, applied in reverse osmosis and nanofiltration to determine whether a raw water is suitable for membrane filtration or needs pretreatment [Taylor and Jacobs 1996].

### Cleaning strategies

Depending on the type of fouling and the type of membrane, various cleaning methods are available: mechanical cleaning, hydraulic cleaning, and chemical cleaning. Mechanical cleaning is only possible in tubular modules where sponge balls can be pushed through the module. Hydraulic cleaning is widely used with hollow-fiber MF and UF membranes. It consists of a backwash with permeate and/or air, i.e. permeate/air is pushed through the membrane from the permeate side inversing the normal flow direction and thereby dislodging the fouling layer from the membrane surface on the feed side. Backwashes are performed every 30 to 60 minutes and take between 1 to 3 minutes [Jacangelo and Buckley 1996]. However, backwashes are usually not succesful in completely restoring the initial permeate flux necessitating chemical cleanings every few weeks to months (Figure 2.5).

For the chemical cleanings, the membrane is flushed or soaked successively with one or more chemicals. Of course, the resistance of the membrane to the chemical used must be assured. Typical chemicals are according to Mulder [1991]:
- acids (e.g. citric acid) to remove scaling,
- bases (mainly NaOH) to remove organic fouling,
- oxidants/ disinfectants (e.g. $H_2O_2$, chlorine) to oxidize organic fouling and inhibit biofouling,
- detergents,
- enzymes, and
- complexing agents (e.g. EDTA).

Ideally, the chemical cleaning should restore the initial flux (and/or initial transmembrane pressure) completely (Figure 2.5, first chemical cleaning). This ensures a viable long-term membrane operation. However, irreversible fouling may occur and diminish the permeate production over time (Figure 2.5, second chemical cleaning).

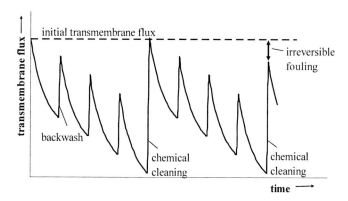

**Figure 2.5**     Restoration of transmembrane flux by backwash and chemical cleaning

Although different cleaning schemes are available (regular backwash with and without airscrub, chemical cleanings), the underlying mechanisms of fouling are not well understood. Instead the cleaning schemes are based on trial-and-error and experience.

### 2.1.5  Fouling mechanisms and models

A number of fouling models exist in order to describe different fouling mechanisms encountered during membrane filtration. For constant pressure filtration, Hermia [1982] presents a physical model for the mechanisms of cake formation, pore constriction, intermediate and complete pore blockage. The general mathematical form is:

$$\frac{d^2t}{dV^2} = k\left(\frac{dt}{dV}\right)^n \qquad \text{(eq. 2.8)}$$

with

| | | |
|---|---|---|
| t | = | filtration time |
| V | = | total filtered volume |
| k | = | rate constant depending on n |
| n | = | filtration constant characterizing the filtration model |

Depending on the number of the exponent n, a different fouling mechanism predominates:
- cake formation for $n = 0$,
- intermediate blocking (pore blockage with particles settling on other particles) for $n = 1$,

- pore constriction or standard blocking for n = 1.5, and
- complete pore blocking (each particle seals a pore; no superimposition of particles on top of one another) for n = 2.

The flux data collected during membrane filtration experiments can be examined for the prevailing fouling mechanism by replotting the data using characteristic coordinates. A linear plot of time/volume versus volume is found in the case of cake formation. If pore constriction is the predominate fouling mechanism, the data is linearized using a plot of time/volume versus time while a linear plot with flux versus volume as coordinates indicates pore blockage.

Oftentimes a clear linear trend cannot be detected on either of these plots and instead a transition from one fouling mechanism to another appears to occur during the filtration. Ho and Zydney [2000] propose a combined pore blockage – cake filtration model for dead-end filtration in order to account for this transition in fouling mechanisms:

$$\frac{Q}{Q_0} = \exp\left(-\frac{\alpha\,\Delta p\,c_b}{\mu\,R_m}t\right) + \frac{R_m}{R_m + R_{pr}}\cdot\left(1 - \exp\left(-\frac{\alpha\,\Delta p\,c_b}{\mu\,R_m}t\right)\right) \qquad \text{(eq. 2.9)}$$

where

$$R_{pr} = \left(R_m + R_{pr0}\right)\sqrt{1 + \frac{2f'R'\Delta p\,c_b}{\mu\left(R_m + R_{pr0}\right)^2}t} - R_m \qquad \text{(eq. 2.10)}$$

with

| | | |
|---|---|---|
| Q | = | volumetric flow rate [m³/s] |
| $Q_0$ | = | initial volumetric flow rate [m³/s] |
| $\alpha$ | = | pore blockage parameter [m²/kg] |
| $\Delta p$ | = | transmembrane pressure difference [N/m²] |
| $c_b$ | = | bulk concentration of protein [g/L] |
| t | = | filtration time [s] |
| $\mu$ | = | viscosity [kg/sm] |
| $R_m$ | = | hydraulic resistance of membrane material [1/m] |
| $R_{pr}$ | = | hydraulic resistance of protein deposit [1/m] |
| $R_{pr0}$ | = | resistance of a single protein aggregate [1/m] |
| R' | = | specific protein layer resistance [1/m] |
| f' | = | fractional amount of total protein that contributes to deposit growth |

The model is developed for protein fouling of microfiltration membranes. According to the authors, the permeate flow rate depends on the rate of pore blockage ($\alpha$), the resistance of a single protein aggregate ($R_{pr0}$), and the increase of the hydraulic resistance of the protein deposit over time (f'·R'). As most membrane filtration systems are operated in a cross-flow mode, Kilduff et al. [2002] have incorporated an additional term for solute back-transport into the bulk solution in this combined pore blockage – cake formation model and applied it to natural organic matter membrane filtration.

## 2.2    Applications of membrane filtration in municipal wastewater treatment

Low-pressure membranes are being used by the wastewater industry either as a polishing step after secondary/tertiary treatment or as part of membrane bio-reactors (MBR). Generally, pressure-driven membrane filtration can be integrated into the wastewater treatment process at two different points: 1) after the secondary clarifiers for effluent polishing and 2) after the biological treatment in the form of MBR systems (Figure 2.6).

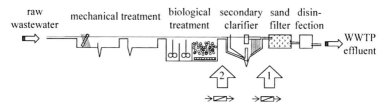

**Figure 2.6**    Possible applications of membrane filtration in wastewater treatment:
1) low-pressure membrane filtration of effluent
2) combination of biology and membrane filtration in MBRs
adapted from [Voßenkaul and Rautenbach 1997]

If the membranes are placed at the end of the treatment process after the secondary clarifiers, they can replace chemical disinfection – mandatory in the United States for all municipal wastewaters – resulting in a higher quality effluent as microorganisms and turbidity are physically retained. In membrane bio-reactors, the membrane replaces the secondary clarifiers and is directly integrated into the biological treatment process ensuring high quality effluents because of the complete biomass retention. The permeate of both applications, effluent membrane filtration and membrane bio-reactors, can be used for wastewater reclamation and reuse applications, e.g. prior to reverse osmosis.

### 2.2.1   Advanced treatment of municipal WWTP effluents with membranes

In light of necessary water conservation, an increasing number of utilities are looking into the possibility of reusing wastewater. For this purpose, secondary or tertiary effluents from municipal wastewater treatment plants (=WWTP) are further treated using membrane filtration. The intended reuse varies, e.g. prevention of seawater intrusion, recharge of groundwater sources for drinking water production or production of process water for industrial use [Reith and Birkenhead 1998, Menkveld et al. 2003].

Although more and more membrane plants are operated for the treatment of wastewater effluent, the challenge posed by fouling remains. Various studies have been investigating different aspects of membrane fouling, especially with respect to fouling control. DeCarolis et al. [2001] assessed the effect of operating conditions such as operational flux, backwash frequency, and in-line coagulation pretreatment on membrane productivity for ultrafiltration of tertiary wastewater for water reuse. Similarly, Parameshwaran et al. [2001] looked at the influence of the permeate flux on filtration cycle time, energy demand, production costs, cake resistance, and fouling during secondary effluent MF membrane filtration. Besides operating conditions such as cross-flow velocity, transmembrane pressure, and backflushing method, Tchobanoglous et al. [1998] determined total solids and particle size distribution as significant parameters to assess membrane performance of ultrafiltration of secondary municipal wastewater effluents.

Roorda and van der Graaf [2003] investigated the filtration characteristics of ultrafiltration of Dutch WWTP effluents by determining the specific ultrafiltration resistance (SUR = product of specific cake resistance of the retained solids and solids concentration). The calculation of the SUR parameter is based on the cake formation model described by Hermia [1982] (see Section 2.1.5). The authors found that the filtration characteristics of ultrafiltration membranes are primarily influenced by effluent material in the size range 0.1-0.2 µm, i.e. 5-20 times the pore diameter of the applied UF membrane, as this fraction is predominant in the formation of a cake layer on the membrane. They conclude that a stable UF membrane performance is only possible under the precondition that the WWTP effluent exhibits a SUR of $\leq 10*10^{12}$ m$^{-2}$. Jarusutthirak et al. [2002] investigated the fouling of nanofiltration and ultrafiltration membranes by wastewater treatment plant effluent organic matter (EfOM) with Fourier transform infra-red spectroscopy (FTIR). The main foulants were identified as polysaccharide-like and/or protein-like material.

## 2.2.2  Membrane bio-reactors in municipal wastewater treatment

Microfiltration and ultrafiltration membranes are gaining importance in membrane bio-reactors where the membranes replace the secondary clarifier in the conventional wastewater treatment process. Three different configurations can be found with regards to the placement of the membrane (Figure 2.7) depending on the type of membrane and the manufacturer. Hollow-fiber membranes with an inside-out flow pattern and plate and frame membrane modules are installed in external pressure vessels and operate in the cross-flow mode (Figure 2.7a). Immersed membranes, i.e. hollow-fiber membrane modules with an outside-in flow pattern, are installed either in a separate filtration tank (Figure 2.7b) or directly in the aerated zone of the activated sludge process (Figure 2.7c). In both cases, the necessary transmembrane pressure is provided through a suction pump on the permeate side of the immersed membranes. An additional pump is needed to ensure the recirculation of the activated sludge suspension if the immersed membrane is installed in a separate tank. The advantage of this configuration (b) is that the membrane does not have to be removed for chemical cleanings.

a)

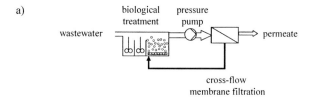

cross-flow
membrane filtration

b)

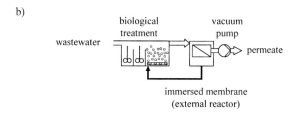

immersed membrane
(external reactor)

c)

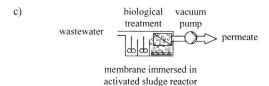

membrane immersed in
activated sludge reactor

**Figure 2.7**     Configurations and placement of the membrane in MBR processes:
a) cross-flow membrane filtration with pressurized membrane reactor
b) immersed membrane with suction pump in separate membrane reactor
c) membrane immersed in aerobic reactor of the activated sludge treatment
adapted from [Voßenkaul and Rautenbach 1997]

World wide, 500 MBRs are operational treating different kinds of wastewater, i.e. industrial wastewater (27 %), domestic wastewater (27 %), in-building greywater (24 %), municipal wastewater (12 %), and landfill leachate (9 %) [Stephenson et al. 2003]. The in-building treatment of greywater with membrane bio-reactors for reuse as toilet flush water is especially common in Japan where these systems have been running since the early 1980s.

The advantages of MBR systems compared to conventional wastewater treatment plants are a high quality effluent with respect to the organic carbon load, and the absence of suspended matter and microorganisms which eliminates the need for additional disinfection [Manem and Sanderson 1996]. Furthermore, the higher achievable biomass concentrations (the sludge retention time is no longer coupled to the hydraulic retention time) allow one to build more compact systems compared to the footprint of conventional activated sludge systems. This makes MBRs especially suited for small scale applications (in-building wastewater reuse, decentralised WWTP).

As with membrane filtration of WWTP effluents, the fouling of low-pressure membranes in MBRs is not well understood. In contrast to effluent membrane filtration, the membranes of MBRs are exposed to the activated sludge. Thus, besides colloidal and dissolved organic matter, suspended solids (i.e., activated sludge flocs and microorganisms) can contribute to membrane fouling. This is supported by some authors who state that the solids fraction impacts the fouling most significantly, e.g. Defrance et al. [2000]. However, the influence of the mixed liquor suspended solids (MLSS) concentration is controversial. An increase in filterability with increasing MLSS concentration is reported [Lee et al. 2003, Shin et al. 2002] as well as the opposite, i.e. a negative impact of increasing MLSS concentration [Sato and Ishii 1991, Madaeni et al. 1999].

According to others, the non-settable organic fraction (i.e. organic colloids and solutes) of the activated sludge induces the most significant membrane fouling in MBRs [Chang et al. 2001, Bouhabila et al. 2001, Ishiguro et al. 1994]. Generally, organic colloids and dissolved organic substances have a negative impact on the membrane performance. However, the relevant compounds are usually identified and quantified using composite parameters such as DOC and UV absorption, e.g. Shin and Kang [2003] use these two parameters as measurement for the SMP (soluble microbial products) production in an MBR fed with glucose. Other researchers use colorimetric methods to quantify polysaccharides (as glucose equivalent) and proteins (as bovine serum albumin equivalent) as main components of extracellular polymeric substances (EPS) [Rosenberger 2003, Evenblij and van der Graaf 2003]. In this research, size exclusion chromatography with on-line UV absorption and organic carbon detectors is employed. This enables the characterisation of the organic matter according to its size and the identification of the substances of the bulk organic matter responsible for membrane fouling.

# Chapter 3     Experimental set-up and analyses

## 3.1     Stirred cell testing

The fouling potential is determined as flux decline over time using Amicon 8200 dead-end stirred cells (Millipore, USA) as shown in Figure 3.1. The cells have a volume of 200 mL and the effective membrane filtration area is 28.7 cm². Due to an attached feed reservoir it is possible to filter samples of up to 4 L. The water reservoir is filled with the sample and pressurized using nitrogen gas (5.0 grade). The sample in the filtration cell is stirred over the entire experiment to minimize concentration polarization and the membrane flux is measured using a volumetric cylinder and stopwatch. All experiments are run at room temperature (~ 22°C) and constant pressure. Experiments are carried out at 0.3 bar and 0.6 bar (MF) or 1.0 bar (UF) depending on the membrane used (see Table 3.1).

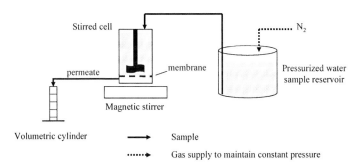

**Figure 3.1**     Experimental set-up for stirred cell testing

A new membrane is used for each experiment unless stated otherwise. Prior to use each membrane is placed in ultra-pure water for at least 24 h to remove wetting agents and production residues. Immediately before the stirred cell test the pure water flux of the membrane is determined by filtering ultra-pure water through the membrane until a stable permeate flux is reached.

In full-scale membrane filtration applications, backwash cycles with and without airscrub and chemical cleaning intervals are incorporated in the operation of the membrane system to maximize efficiency, i.e. permeate flux versus applied pressure. To simulate backwash cylces in the stirred cell tests, hydraulic cleaning schemes are carried out in some of the experiments. The hydraulic cleaning scheme consists of i) emptying the stirred cell of the water sample, ii) turning the membrane upside-down, iii) filtering 180 mL ultra-pure water or permeate through the reversed membrane at the same transmembrane pressure as applied during filtration (=backwash without airscrub), iv) turning the membrane right-side up, v) refilling the stirred cell with the water sample and resuming filtration.

In total five different membranes are used to test a range of properties including materials, pore sizes/ molecular weight cut-offs, and contact angles (see Table 3.1). Two microfiltration membranes with a nominal pore size of 0.22 µm are selected to assess the influence of hydrophobicity on the fouling behaviour. The GVHP membrane is made of polyvinylidene fluoride (PVDF) and exhibits an hydrophobic character as indicated by a contact angle of 83°. Membranes with contact angles above 50° are considered to be hydrophobic, those with contact angles below 50° are hydrophilic. The hydrophilic membrane (GWSP, contact angle 19°) is manufactured from a mixture of cellulose acetate and cellulose nitrate (CA/CN). Experiments with this membrane are carried out at 0.3 bar, while a pressure of 0.6 bar is applied during membrane filtration with the hydrophobic membrane (GVHP) in order to obtain sufficient flux.

The two tighter membranes (MX500 and YM100) are chosen to be able to discern between different fouling mechanisms such as cake formation, pore blockage, and adsorption. The MX500 membrane lies on the border of micro- and ultrafiltration with a nominal pore size of 0.05 µm. The membrane material is polyacrylonitrile (PAN). As a loose ultrafiltration membrane, the YM100 membrane has a nominal molecular weight cut-off (MWCO) of 100 000 Dalton (D) and is made of regenerated cellulose. Both membranes are hydrophilic (4° and 18°, respectively) and stirred cell experiments are carried out at 1 bar.

The fifth membrane (VVLP) is chosen due to its similarity to the membrane used in the pilot plants in Berlin-Ruhleben, Germany. The pore size lies between 0.1 µm and 0.2 µm and the membrane material is polyvinylidene fluoride (PVDF). Contrary to the GVHP membrane, however, this membrane is hydrophilized during its fabrication. The applied pressure during the stirred cell experiments is 0.3 bar. This corresponds to the transmembrane pressure (TMP) threshold in the pilot plants which are operated under constant flux conditions, i.e. when a TMP of 0.3 bar is reached a chemical cleaning of the membrane is initiated.

**Table 3.1**        Characteristics of flatsheet membranes used in stirred cell units

| Membrane | Material | Pore size MWCO | Contact angle, hydrophobicity | Manufacturer | applied pressure |
|---|---|---|---|---|---|
| GVHP | PVDF | 0.22 µm | 83°, hydrophobic | Millipore | 0.6 bar |
| GSWP | CA/CN | 0.22 µm | 19°, hydrophilic | Millipore | 0.3 bar |
| VVLP | PVDF | 0.1-0.2 µm | hydrophilized | Millipore | 0.3 bar |
| MX500 | PAN | 0.05 µm | 4°, hydrophilic | Osmonics | 1 bar |
| YM100 | RC | 100 000 D | 18°, hydrophilic | Amicon | 1 bar |

PVDF: Polyvinylidene fluoride          CA/CN: mixture of cellulose acetate and cellulose nitrate
PAN: Polyacrylonitrile          RC: regenerated cellulose

Flux decline results from the stirred cell experiments are shown as normalized flux $J/J_0$ over delivered DOC. $J_0$ is the initial flux for the water sample as measured after the first minute of filtration. This definition of $J_0$ is chosen instead of using the ultra-pure water flux as $J_0$ because it allows for a better comparison of different flux decline curves, i.e. all curves start at $J/J_0=1$. The delivered DOC is the cumulative dissolved organic carbon delivered to the membrane surface over the experiment. It is proportional to the initial dissolved organic carbon of the feed water and the filtered permeate volume:

delivered DOC [mg C/m²] = $DOC_{feed}$ [mg/L] * permeate volume [L] / membrane area [m²].

## 3.2    Membrane bio-reactor pilot plants

Two membrane bio-reactor (MBR) pilot plants are operated at the wastewater treatment plant Berlin-Ruhleben (WWTP Ruhleben) by Veolia Water and Berliner Wasserbetriebe. Both plants are fed with degritted raw water as it is treated in the conventional wastewater treatment plant. The primary aim of the piloting is the investigation of biological phosphorus removal in conjunction with nitrification/denitrification in MBRs for later use in remote areas and small scale applications (WWTP serving a few thousand inhabitants) [Gnirss et al. 2003a].

The two pilot plants are equipped with a hollow-fiber module from USF Memcor, Australia, made of polyvinylidene fluoride (PVDF) and a pore size of 0.1 – 0.2 μm. The operating conditions (e.g., loading rate, hydraulic retention time, sludge retention time, temperature…) are identical for the two pilot plants. The pilot plants have been operated since September 2001 and have been operated under different sludge retention times (SRT) from 8 to 26 days [Gnirss et al. 2003a]. All experiments carried out for this research have been performed between August 2002 and November 2003.

The difference between the two pilot plants is the placement of the anoxic zone (see Figure 3.2): pilot plant 1 (PP 1) is operated in the conventional pre-denitrification mode where the aerobic zone is preceded by the anoxic zone, while the aerobic zone is followed by the anoxic zone in pilot plant 2 (PP 2). Thus, pilot plant 2 operates in a post-denitrification mode without additional dosing of a carbon source.

**Pilot Plant 1**

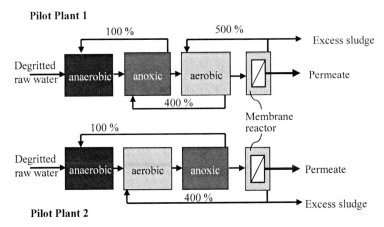

**Figure 3.2**    Schematic of the two MBR pilot plants operated at the municipal wastewater treatment plant Berlin-Ruhleben, Germany

During this study, the two pilot plants are operated at two constant sludge retention times (SRT): eight days from January 2003 – May 2003 and fifteen days from July 2003 – November 2003. The contact time in the membrane reactor is four minutes in both pilot plants. Due to the different configurations, the two pilot plants have slightly different volumes: 2.1 m³ for pilot plant 1 (pre-denitrification) and 1.9 m³ for pilot plant 2 (post-denitrification). The operating conditions for the two sludge retention times and both pilot plants are summarized in Table 3.2. On 29th July 2003 the module A3 (flat sheet membrane) has been connected in by-pass to pilot plant 1. The reactor volume and throughflow are kept proportional and therefore, the load for the Memcor hollow-fiber membrane has remained identical.

**Table 3.2**    Operating conditions for MBR pilot plants at SRT of 8 d and 15 d

|  | Pilot plant 1 | | Pilot plant 2 | |
| --- | --- | --- | --- | --- |
| SRT | 8 d | 15 d | 8 d | 15 d |
| HRT (h) | 11.4-12.4 | 9.9-12.4 | 12.1-12.6 | 12.3-14.4 |
| Throughflow (L/h) | 182.7-203.2 | 180.3-313* | 161.2-174.5 | 128.2-167.4 |
| Permeate flux (L/hm²) | 22.5-25.0 | 18.0-22.5 | 20.9-21.6 | 14.9-21.6 |
| MLSS (g/L) | 7.3-12.9 | 9.2-16.5 | 7.1-11.3 | 10.4-14.7 |

* includes flow through A3 module from 29th July 2003: 107-129 L/h

For both sludge retention times (SRT of 8 d and 15 d) the throughflow, permeate flux, and MLSS are slightly higher in pilot plant 1 than in pilot plant 2 while the hydraulic retention time is longer in PP 2. Nevertheless, all parameters are in the same range for the two pilot plants. For further information on operating conditions and results of the MBR see Gnirss et al. [2003b].

## 3.3    Feed water for stirred cell testing

Two wastewater treatment plants (WWTP) are investigated in this research: one from the United States (WWTP Boulder, Colorado) and one from Germany (WWTP Ruhleben, Berlin).

### 3.3.1  WWTP Ruhleben

**Conventional wastewater treatment plant**

The wastewater treatment plant Ruhleben is the largest of five plants treating municipal wastewater from Berlin, Germany. Raw wastewater, including 20-30 % industrial wastewater, is treated by conventional activated sludge process with biological phosphorous removal and biological nitrification/denitrification (using a pre-denitrification configuration). The plant's capacity is 240,000 m³/d and its annual average effluent quality is summarized in Table 3.3.

**Table 3.3**    Annual average values of the conventional effluent of the WWTP Ruhleben for 2002 [Gnirss 2004, Gnirss et al. 2003b]

| COD | BOD | TOC | AOX | TKN | $NH_4^+ - N$ | $NO_3^- - N$ | TP | $Cl^-$ |
|------|------|------|------|------|------|------|------|------|
| mg/L | mg/L | mg/L | µg/L | mg/L | mg/L | mg/L | mg/L | mg/L |
| 43 | 3.7 | 14 | 76.5 | 12 | < 0.4 | 7.7 | 0.3 | 142 |

Two different kind of samples are used (grab samples):
i) Activated sludge from the conventional WWTP which is filtered over paper filter (589/1 black ribbon[1], Schleicher & Schuell GmbH, Germany) to obtain the aqueous phase only. These samples are referred to in this text as "filtrate CAS" (conventional activated sludge). ii) Effluent from the conventional WWTP which is referred to as "Ruhleben effluent" or "Ruhleben EfOM".

### 3.3.2  MBR pilot plants

The two pilot plants are described in Section 3.2. Samples are taken from the membrane reactors and filtered over paper filter (black ribbon, Schleicher & Schuell GmbH, Germany) in order to separate the sludge from the water phase. The black ribbon paper filters are used in accordance to the German standard method for sludge filtration (DIN 38409 part 2). The paper filters are rinsed with 200 mL permeate. The filtered activated sludge from the membrane reactors is referred to as "filtrate PP 1" or "filtrate PP 2", depending on the pilot plant sampled. To minimize variations due to daily and/or weekly devolution, samples are taken on the same weekday between 8 am and 9 am, seven minutes after the last backwash (a backwash is performed every 12 min). Additionally, permeate grab samples are used for experiments (= "permeate PP 1" or " permeate PP 2").

---

[1] retention range: >12-25 µm

### 3.3.3  WWTP Boulder

**Conventional wastewater treatment plant**

The City of Boulder wastewater treatment plant, Colorado, treats a maximum of 20.5 MGD ($\approx$ 77600 m³/d) municipal wastewater with trickling filters followed by activated sludge (operated in non-nitrifying mode) as this ensures more stable operating conditions than with either trickling filters or activated sludge alone. One third of the wastewater is then treated for ammonia removal (nitrification over an additional trickling filter) and recombined with the other two thirds before chlorination/dechlorination. Contrary to Germany, wastewater treatment plant effluents are disinfected in the United States before discharge. In the case of chlorination, a de-chlorination step is needed to prevent chlorine from entering the receiving water body where it would kill aquatic life. The annual average effluent quality is summarized in Table 3.4. Samples used for this research are grab samples from the discharged effluent (after chlorination/ dechlorination) which are filtered over 0.45 μm directly after sampling. These samples are referred to as "Boulder effluent" or "Boulder EfOM".

**Table 3.4**    Annual average values of the conventional effluent of the WWTP Boulder for 2001 [City of Boulder 2002]

| BOD | TKN | $NH_4^+ - N$ | $NO_3^- - N$ |
|------|------|------|------|
| mg/L | mg/L | mg/L | mg/L |
| 13.0 | 10.1 | 7.6 | 9.0 |

**Isolates**

In addition to bulk wastewater, fractions of the organic carbon present in wastewater treatment plant effluent are studied. The objective is to determine the contributions of all fractions to the fouling of low-pressure membranes. Effluent from the City of Boulder wastewater treatment plant, Colorado, USA, is used to fractionate and isolate effluent organic matter (EfOM). Table 3.5 gives an overview of the different isolates used in stirred cell testing. Information on the isolation procedure are given in Section 3.4. For each stirred cell test experiment, 20 mg of the corresponding lyophilized isolate are re-dissolved in 2 L ultra-pure water. This yields a dissolved organic carbon content of approximately 4 to 5 mg/L. The pH is adjusted with sodium hydroxide (NaOH) to 6.8 and the conductivity with KCl to 700 μS/cm to match the conductivity of Boulder effluent.

**Table 3.5**     Isolates used in experiments to determine fouling potential and behaviour

| Label | Sample characteristics |
|---|---|
| colloids > 6-8 kD | Organic colloids from dialysis bag with MWCO 6-8 kD |
| colloids >12-14 kD | Organic colloids from dialysis bag with MWCO 12-14 kD |
| HPO-A | Hydrophobic acids back-eluted from XAD-8 resin with NaOH |
| TPI-A | Transphilic acids back-eluted from XAD-4 resin with NaOH |

## 3.4    Isolation procedures

Colloids are isolated from a 100 L wastewater effluent sample. After prefiltration through 0.7 μm Balston glass fiber filters (Parker Hannifin Corporation, Ohio, USA) the sample is vacuum evaporated (rotary-evaporator Büchi, Brinkmann, New York, USA) to 100 mL. In order to redissolve precipitating calcium carbonate 1 N HCl is added during the process (3x 100 mL). The obtained slurry is put in a dialysis bag (Spectra/Por 3) with a MWCO of 3500 Dalton and dialysed against ultra-pure water until the conductivity of the dialysis permeate equals that of ultra-pure water, i.e. below 10 μS/cm. During the dialysis all ions, salts, and molecules with a molecular weight of less than 3500 D are separated from the colloids. Everything that remains in the dialysis bag at the end of the dialysis is freeze-dried (lyophilization) and stored for later use in stirred cell testing. In the following these colloids are referred to as "colloids > 3500 D" (see Figure 3.3).

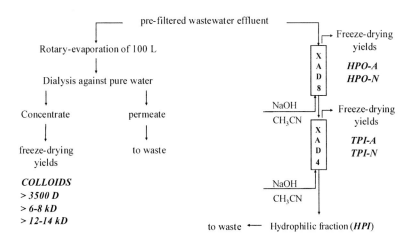

**Figure 3.3**     EfOM isolation and fractionation scheme: i) isolation of colloids includes rotary-evaporation followed by dialysis (concentrate=colloids) and freeze-drying of the isolates; ii) isolation of hydrophobic and transphilic organic material without previous removal of colloids consists of XAD-8/ XAD-4 resins (XAD-8=HPO-A, HPO-N; XAD-4=TPI-A, TPI-N) and freeze-drying

Size exclusion chromatography of the isolate "colloids > 3500 D" gives the organic carbon chromatogram in Figure 3.4a. Apparently, some smaller organic molecules remain in the dialysis bag together with the organic colloids. In order to optimize the separation of low-molecular weight material from the colloids, one third of the slurry is transferred from the 3500 D dialysis bag into a dialysis bag with a MWCO of 6-8 kD and another third of the slurry into a dialysis bag with a MWCO of 12-14 kD. The corresponding isolates are referred to as "colloids > 6-8 kD" and "colloids > 12-14 kD", respectively. Chromatograms of these two isolates are depicted in Figure 3.4b and do not exhibit the smaller molecular weight peak after 3700 sec retention time.

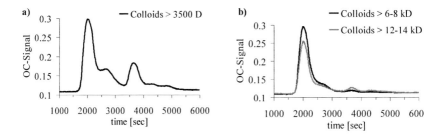

**Figure 3.4**     SEC chromatograms (organic carbon response) of colloidal isolates:
a) colloids > 3500 D
b) colloids > 6-8 kD (top) and colloids > 12-14 kD (bottom)

Furthermore, Amberlite XAD-8 and XAD-4 resins are used to isolate hydrophilic organic matter following the method of Aiken et al. [1992]. The Amberlite XAD resins are macroporous non-ionic resins made of acrylic ester (XAD-8) and styrene divinylbenzene (XAD-4), respectively. Originally, the method has been developed to isolate different fractions of natural organic matter (NOM) as present in surface waters. The idea was to be better able to characterise the different components or compound classes of natural organic matter, e.g. by elemental analysis, molecular weight determination, acid-base titration, amino acid analysis, [13]C-NMR, [1]H-NMR, and/or infra-red (IR) spectroscopy [Aiken and Leenheer 1993]. The primary interest was to isolate aquatic humic substances from the environment in order to define their structure and chemical properties. These informations are used to further the understanding of the role aquatic humic substances play in the environment as well as their biogenesis [Aiken 1988].

With this method the organic matter of a water sample is fractionated into hydrophobic acids (HPO-A) and hydrophobic neutrals (HPO-N) adsorbing onto the XAD-8 resin, transphilic acids (TPI-A) and transphilic neutrals (TPI-N) adsorbing onto the XAD-4 resin, and the hydrophilic fraction (HPI) which does not adsorb on either the XAD-8 or XAD-4 resin and is disposed off (Figure 3.3). However, the separation of the different organic matter fractions with XAD resins is not sharp, instead the fractions overlap to a certain degree [Aiken 1988, Aiken and Leenheer 1993]. It should be mentioned that the so-called 'hydrophobic' fractions (adsorbing onto the XAD-8 resin) do not exhibit a truly hydrophobic character in the chemical sense. The organic matter found in the 'hydrophobic' fraction is not hydrophobic such as lipids are hydrophobic. They merely exhibit a more hydrophobic character in comparison to the transphilic and the hydrophilic fractions (sometimes also-called hydrophilic and ultra-hydrophilic fractions). All five fractions are hydrophilic in the chemical sense as the bulk organic matter is present in the original water sample in the form of dissolved and colloidal macromolecules. In this approach (XAD-8 followed by XAD-4), most of the colloids end up in the hydrophilic fraction (HPI), although, small amounts of colloids may be included in the other two fractions due to physical filtration by the resins.

The method consists of a pre-filtration step (0.7 µm Balston glass fiber filters), the acidification of the sample with hydrochloric acid (concentrated HCl) to pH < 2, and the consecutive loading of an sample aliquot (102 L of Boulder WWTP effluent) onto a 2 L XAD-8 column followed by a 2 L XAD-4 column. After the loading is completed the columns are back-eluted separately with 4 L of 0.1 N NaOH each in order to obtain the acids fractions (HPO-A and TPI-A). The hydrophobic acids fraction and the transphilic acids fraction are then passed through cation exchange columns and freeze-dried. To desorb the neutral fractions (HPO-N and TPI-N) the resins are back-eluted with an acetonitrile-water mix (75 % $CNCH_3$ and 25 % $H_2O$) and the acetonitrile is rotary-evaporated off before lyophilization. A total volume of 1530 L Boulder WWTP effluent is processed according to this procedure. Only the hydrophobic acids and the transphilic acids fractions are used in this research.

## 3.5 Size exclusion chromatography

### 3.5.1 Size exclusion chromatography systems

Size exclusion chromatography (SEC) with UV and online organic cabon detection is used for the characterisation of the organic carbon in the samples. Two different SEC systems are used for this research, as part of it is done in Boulder, Colorado, USA, and part of it in Berlin, Germany. To facilitate the identification of each system, the system in Boulder will be referred to as HPLC-SEC in this text and the system in Berlin as LC-OCD (as it is called by the manufacturer DOC-LABOR Dr. Huber, Karlsruhe, Germany). LC-OCD stands for Liquid Chromatography – Organic Carbon Detection.

Both systems consist of a size exclusion chromatography column in order to separate organic molecules according to their molecular size. The underlying principle is the diffusion of molecules into the resin pores. This means that larger molecules elute first as they can not penetrate the pores very deeply, while smaller molecules take more time to diffuse into the pores and out again.

Table 3.6 shows the characteristics of the two columns (columns are 250 mm long and have a diameter of 20 mm) used for this research, such as pore size of resin and range of molecular weight separation. The resin consists of semi-rigid, spherical beads with a hydrophilic surface and is synthezised by co-polymerisation of ethylene glycol and methacrylate-type polymers. The supplier (GROM Analytik + HPLC GmbH, Herrenberg, Germany) states the molecular size separation range detectable according to polyethylene glycols (PEG), dextrans, and globular proteins as standard molecules. The SEC systems are operated with either the column HW-50S (Boulder,USA) or with the HW-55S column (Berlin, Germany).

**Table 3.6**     Columns used in the LC-OCD system at the Technical University Berlin (TU) and in the HPLC-SEC system at the University of Colorado at Boulder (CU); characteristics according to the supplier GROM Analytik + HPLC GmbH [GROM 2003]

| Toyopearl TSK | particle size [μm] | pore size [Å] | PEG [D] | Dextran [D] | Globular Proteins [D] |
|---|---|---|---|---|---|
| HW-50S | 20 - 40 | 125 | 100 – 18 000 | 500 – 20 000 | 500 – 80 000 |
| HW-55S | 20 - 40 | 300 | 100 – 150 000 | 1 000 – 200 000 | 1 000 – 700 000 |

Besides the different columns used in the two systems, the operating conditions are similar (flowrate: 1 mL/min, eluent: phosphate buffer) although the sample preparation is not. For the system in Boulder (HPLC-SEC) 2 mL of the sample are injected after adjusting the pH and the conductivity of the sample to 6.8 and 5.1 mS/cm (matching the pH and ionic strength of the eluent), respectively. The ionic strength of the eluent is 0.1 mol/L (phosphate buffer with additional sodium sulphate).

In the system in Berlin (LC-OCD), samples are not adjusted to the pH and ionic strength of the eluent (I=0.18 mol/L, pH=6.6, κ=3.1 μS/cm). Additionally, the organic carbon content of the bulk sample is measured at the beginning of each run (injection at 0 minutes) using a by-pass around the column in the LC-OCD set-up. This gives the by-pass peak. After approximately 10 minutes the sample is injected into the column.

Figure 3.5 provides an overview of the two system set-ups. The separated compounds are detected by UV absorption at 254 nm (Berlin: WellChrom fixed wavelength detector K-200, Knauer, Berlin, Germany; Boulder: SPD-6A UV Spectrophotometric detector, Shimadzu, Columbia, USA) followed by dissolved organic carbon (DOC) detection. In order to eliminate any inorganic carbon, phosphoric acid is added after the UV detector in both systems. In the LC-OCD (see Figure 3.5a) the organic carbon is oxidised in the Gräntzel thin-film reactor by radiolytically produced oxygen radicals (aqueous sample + 185 nm UV light under nitrogen atmosphere [Huber and Frimmel 1991]). The produced carbon dioxide is detected by non-dispersive infra-red absorption (Ultramat 6, Siemens, Munich, Germany). In the HPLC-SEC (Boulder) the organic carbon is oxidised and measured in a modified Sievers 800 Turbo portable total organic carbon analyser (Ionics Instruments Business Group, Boulder, USA) by UV/persulfate oxidation and conductivity detection of the resulting carbon dioxide as carbonates [Her et al. 2002].

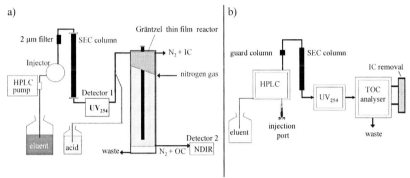

**Figure 3.5**    Set-up of the two size exclusion chromatography sytems: a) LC-OCD (TU Berlin), b) HPLC-SEC (CU Boulder); both systems have automated data acquisition for the detectors (not shown)

The differences in operation lead to two major differences in the resulting chromatograms:
1) HPLC-SEC (Boulder) chromatograms do not exhibit an acid peak as found in the LC-OCD (Berlin) chromatograms. This is due to the pH and conductivity adjustment of the samples measured in Boulder, USA.
2) Because of the TOC by-pass measurement by the LC-OCD, the first 10 min (by-pass peak) do not pertain to the chromatogram itself. These first ten minutes are, however, not substracted from the chromatograms in this study. This is legitimate as by-pass measurements are performed on both calibration standards and samples, and only chromatograms from the same system are directly compared.

### 3.5.2   Analysis of chromatograms

In ideal chromatography, the molecules are retained due to diffusion only. Very hydrophobic or neutral molecules, however, may interact with the resin. Due to these non-ideal chromatographic effects, these molecules do not elute according to their size. Instead they elute after long retention times, i.e. after the salt boundary. Hence, no conclusions may be drawn about the molecular size of such hydrophobic substances.

Figure 3.6 shows the relevant boundaries: (i) the void volume peak which can be measured using Blue Dextran 2000 with a molecular weight of ~ 2,000,000 Dalton, and (ii) the salt boundary which is determined with sodium nitrate ($NaNO_3$) and the UV detector. Within these boundaries one can assume a more or less ideal size exclusion chromatography.

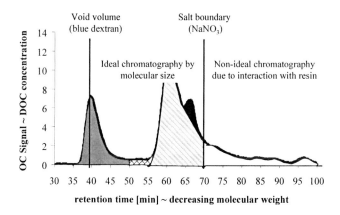

**Figure 3.6**     Ideal versus non-ideal chromatography as defined through the void volume using Blue Dextran 2000 and the salt boundary using NaNO₃

Figures 3.7 and 3.8 show the organic carbon (OC) chromatograms of a typical WWTP effluent sample using a 50S column (HPLC-SEC) and a 55S column (LC-OCD), respectively. The first peak of the organic carbon chromatogram on column 50S (Figure 3.7) after 1900 seconds (peak maximum) is the so-called polysaccharides peak (= PS). Besides polysaccharides, proteins and organic colloids elute in this peak too. It is followed by the humic substances peak (HS) at 3000 seconds which often shows a more or less distinct shoulder/peak attributed to humic hydrolysates or building blocks (BB) around 3400 seconds.

The first peak of the organic carbon chromatogram using column 55S (LC-OCD, Berlin) after 40 minutes (peak maximum) is the polysaccharides peak (= PS). Contrary to column 50S, only organic colloids would also elute in this peak as the molecular size separation spans a wider range (see Table 3.6). This allows for a separation of proteins, e.g. the standard protein bovin serum albumin (BSA) elutes after 52.9 minutes. However, the Ruhleben effluent does not exhibit a distinct peak in this area. Instead the next distinct peak is the humic substances peak (HS). It includes the humic hydrolysates as resolution is lost in the lower molecular weight range in exchange for a better resolution in the larger molecular size range (see above). The third distinct peak in this chromatogram is the acids peak. The acids peak is due to the way the system is operated (non buffered samples) and contains organic acids. After this distinct acids peak, neutral and amphiphilic compounds may show. Nevertheless, it should be kept in mind that anything eluting after the salt boundary at approximately 70 minutes, is interacting with the column resin.

The UV chromatograms show similar distributions with one exception: polysaccharides are not detectable with UV because they do not have aromatic/double bounds necessary for the absorption of light with a wavelength of 254 nm. Nevertheless, a UV absorption peak is often found slightly before the PS peak in the OC chromatogram. This may be due to UV light refraction by organic colloids present in the sample or, as suggested by Huber and Frimmel [1996], light scattering due to inorganic colloids such as silica, clay, and metal oxide (iron or aluminium) colloids.

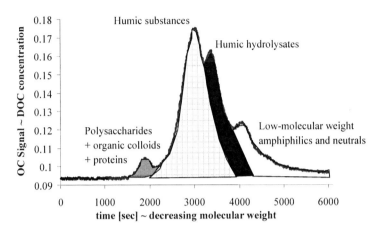

**Figure 3.7**    Attribution of different organic compounds to the main peaks detected during size exclusion chromatography, HPLC-SEC system, column 50S

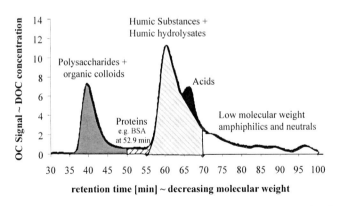

**Figure 3.8**    Attribution of different organic compounds to the main peaks detected during size exclusion chromatography, LC-OCD system, column 55S

### 3.5.3   Calibration of OC detector with potassium phthalate

A calibration of the organic carbon detector is mandatory in order to be able to calculate
the organic carbon content from the peak area. However, this is only done for the LC-OCD
and not for the HPLC-SEC system as the latter one does not have the possibility to by-pass
the columns in order to calibrate the TOC detector. Potassium hydrogen phthalate
($KHC_8H_4O_4$) is used as organic carbon source. All samples are measured using the by-pass
mode since this calibration aims at the infra-red detector itself and how sensitive the $CO_2$
detection is. Figure 3.9 depicts the resulting calibration curve using injection volumes of
50 µL, 100 µL, and 200 µL. The resulting equation to convert peak area into ng C is:
$y = 0.0432x + 0.3383$ ($R^2 = 0.9986$).

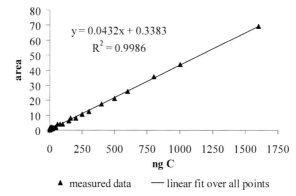

**Figure 3.9**   Calibration curve for organic carbon detector as measured in the by-pass
mode with potassium hydrogen phthalate (injection volumes: 50 µL,
100 µL, and 200 µL)

Regarding the calibration of the UV detector, the DOC-LABOR Dr. Huber (manufacturer)
has calculated a conversion factor of 0.0554 for the LC-OCD system at TU Berlin (for a
sample injection volume of 2000µL). As standards, Suwannee River Humic and Fulvic
Acid (IHSS HA and FA) are used or, more precisely, the ratio of $UVA_{254}$ and OC for each
of them: $UVA_{254}/OC$ (FA) = 4.56 L/(mg*m) and $UVA_{254}/OC$ (HA) = 7.85 L/(mg*m).
These values have been cross-checked with other analysis methods by the DOC-LABOR
Dr. Huber.

### 3.5.4   Calibration of SEC columns with known substances

In order to get an approximate idea of the retention time of various molecular
weights/sizes, a calibration with polyethylene glycols as well as with dextrans is made.
Figure 3.10 shows the OC chromatograms of ten polyethylene glycols ranging from 194
Dalton to 182,000 Dalton for the LC-OCD system (Berlin, Germany). The differing height
of the peaks is due to different organic carbon concentrations of the standards. Of more

interest is the fact that all PEGs elute according to their molecular weight and size. A similar picture is found with dextrans ranging from 1080 D to 123,600 D (data not shown).

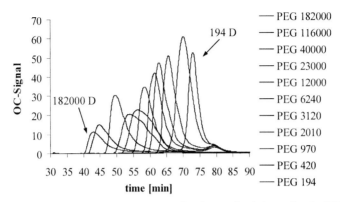

**Figure 3.10**   Superposed chromatograms of various polyethylene glycols (PEGs), from left to right (graph) and top to bottom (legend): large to small

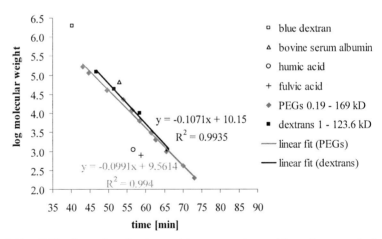

**Figure 3.11**   Calibration of molecular weight versus time with various standard compounds

In Figure 3.11 the peak maximum of each PEG and dextran standard is plotted against its molecular weight. Additionally, the retention time for Blue Dextran 2000 ($\sim$ 2 million D), bovine serum albumin ($\sim$ 67,000 D), humic acid and fulvic acid are shown. The peak maximum of Blue Dextran 2000 elutes after 40 minutes and determines the void volume.

The diagram makes it clear that a calibration with PEGs is not absolute since other molecular structures behave differently, e.g. albumin elutes after the 40,000 D PEG although it has a higher molecular weight. The obtained calibration fits, however, the separation range given by the manufacturer of the columns (see Table 3.6).

## 3.6  Analytical measurements

### 3.6.1  Attenuated total reflectance – Fourier transform infra-red spectroscopy (ATR-FTIR)

Fourier transform infra-red (FTIR) spectroscopy is useful in detecting representative functional groups in organic samples. However, an identification of specific substances is not possible due to the complex matrix of effluent organic matter. Attenuated total reflectance – Fourier transform infra-red spectra are measured using a Nicolet Magna-IR 750 series II FTIR spectrometer from Nicolet, Madison (Wisconsin, USA) with a 45° ZnSe flat plate crystal. The system is set to use an IR light source, a KBr beam splitter, and a DTGS KBr detector. During the spectrum acquisition the measuring chamber is purged with purified air (without $CO_2$, and $H_2O$) in order to minimize interferences by $CO_2$ and $H_2O$. Lyophilized isolates (~ 1 mg) are mixed with KBr powder (~ 1 g) and pressed into pellets prior to analysis. The resulting absorption bands and peaks at different wave numbers can be attributed to specific chemical bonds [Skoog et al. 1998].

For example, polysaccharides are recognized by peaks at wave numbers of 1040 $cm^{-1}$ (C-O bond from alcohol group), 2940 $cm^{-1}$ (-CH stretch), and 3400 $cm^{-1}$ (-OH stretch). A peak at 1040 $cm^{-1}$ could also be indicative for silica although no peak at 2940 $cm^{-1}$ should be detected in this case [Jarusutthirak 2002]. Amide groups which are typical for proteins (peptide bond) absorb infra-red light at wave numbers of 1550 $cm^{-1}$ (C-N bond) and 1640 $cm^{-1}$ (C=O bond). The complexity of humic substances results in various peaks, e.g. at wave numbers of 1620 $cm^{-1}$ (aromatic groups), 1720 $cm^{-1}$ (carboxylic groups), and 3400 $cm^{-1}$ (alcohol groups) [Cho 1998].

### 3.6.2  Total organic carbon

The total organic carbon (TOC) content is measured with either a Sievers 800 portable TOC analyzer (Ionics Instruments Business Group, Boulder, CO, USA) or a high-TOC (Elementar Analysensysteme, Hanau, Germany). The Sievers 800 combines UV/persulfate oxidation of the organic carbon with membrane conductometric detection of the resulting $CO_2$ as carbonates. The high-TOC instrument uses the combustion method to oxidize the organic carbon prior to non-dispersive infra-red detection of the $CO_2$. In both instruments, the samples are acidified in order to purge any inorganic carbon prior to the oxidation of the organic carbon. The dissolved organic carbon (DOC) content is operationally defined by 0.45 μm filtration of the sample prior to analysis.

### 3.6.3  Photometric measurements

*Turbidity*

The turbidity of all pilot plant samples is measured with a laboratory 2100 N turbidimeter from the HACH company, Loveland (Colorado, USA). The calibration of the instrument is carried out with a Gelex® secondary turbidity standard kit. For each measurement, the turbidity cuvette is rinsed with ultra-pure water followed by a rinse with the sample. The analysis is made with 40 mL of sample.

*UV absorption at 254 nm*

The UV absorption at a wavelength of 254 nm ($UVA_{254}$) is measured with either a UV 160U UV/VIS spectrophotometer from Shimadzu, Columbia (MD, USA) for samples analysed in Boulder, USA, or with a Lambda 12 UV/VIS spectrometer from Perkin Elmer, Überlingen (Germany) for samples analysed in Berlin, Germany. In both cases Hellma 10 mm precision cuvettes made of quarz glass Suprasil® are used. The specific UV absorbance at 254 nm (SUVA) is calculated by dividing the $UVA_{254}$ value by the corresponding DOC value.

*Polysaccharide and protein analyses*

The photometric analyses for polysaccharides and proteins and the EPS extraction are carried out by Dr. Sandra Rosenberger according to the procedures explained in Rosenberger [2003]. The photometric analysis of polysaccharides follows the method of Dubois et al. [1956]. The method of Lowry et al. [1951], as modified by Frolund et al. [1996], is used for the photometric analysis of proteins. The extraction of bound extracellular polymeric substances (EPS) from the cell surface is performed with a cation exchange resin (DOWEX) according to the method of Frolund et al. [1996]. The principle is that by exchanging the calcium cations ($Ca^{2+}$), acting as bridges between cell surface and EPS, the EPS can be more easily brought into solution by applying sheer forces (e.g. centrifugation). The activated sludge sample is diluted to 10 g suspended solids /L (the washing step has been omitted). Approximately 70-75 g ion exchanger are needed per g of organic suspended solids. After two hours extraction time, the extracted EPS is separated from the solid phase by centrifugation [Rosenberger 2003].

# Chapter 4    Fouling potential of wastewater treatment plant effluents

To assess the fouling potential of wastewater treatment plant (WWTP) effluents on low-pressure membranes, two different effluents are investigated: Boulder effluent (Colorado, USA) and Ruhleben effluent (Berlin, Germany). A description of the wastewater treatment plants and mean values for the principal parameters (BOD, $NH_3$-N...) are given in Section 3.3. Stirred cell tests are carried out to determine flux decline curves as explained in Section 3.1. Two fouling indices are extracted from the flux decline curves: i) % of initial flux (=$J/J_0*100$) after 500 mg C/m² are delivered to the membrane surface and ii) delivered DOC to the membrane surface (in mg C/m²) at 25 % flux decline. In this chapter, the results of the stirred cell experiments are discussed with regards to the influence of the effluent organic matter (EfOM) on low-pressure membrane fouling. Some of the data in this chapter have been presented at conferences [Reichenbach et al. 2001, Laabs et al. 2002].

## 4.1    Comparison of the fouling behaviour between WWTP effluents

A comparison of the two WWTP effluents is given in Figure 4.1 for the hydrophilic microfiltration membrane (GSWP) and the hydrophilic ultrafiltration membrane (YM100). Both effluents exhibit similar fouling behaviour on each membrane with the Boulder EfOM fouling the membranes to a slightly lesser extent (Table 4.1). Interestingly, the fouling behaviour is also nearly identical between the two membranes although, one is a microfiltration membrane (pore size 0.22 µm) and the other one an ultrafiltration membrane (MWCO 100,000 D). Nevertheless, different fouling mechanisms are possible. With both membranes, most of the permeate flux is lost by the time 1000 mg C/m² DOC are delivered to the membrane surface, i.e. the flux declines to 20-30 % of its initial value. Another 1000 mg C/m² of delivered DOC (= 2000 mg C/m² in total) result only in a further 10 % flux loss. Thus, in a full scale application a backwash would be necessary after 1000 mg C/m² of DOC are delivered to the membrane surface in order to maintain sufficient transmembrane flux. This corresponds to a cumulative filtered volume of 110 L/m² and a filtration time of approximately 15 min in the stirred cell experiments with the MF membrane (GSWP). Jacangelo and Buckley [1996] state that backwashes of MF hollow-fiber membranes are typically performed every 30-60 min for 1-3 min in drinking water applications. However, the filtration cycle between two backwashes can be much shorter in membrane filtration of wastewater. For example, backwashes have been performed every 12 min in the membrane bio-reactor pilot plants monitored during this research (see Section 3.3.2).

The organic compounds present in WWTP effluents that are responsible for fouling are most likely a heterogeneous mixture of large macromolecules with molecular sizes of more than 100,000 D (equals MWCO of the ultrafiltration membrane) up to several hundred nanometers (pore size of the microfiltration membrane is 220 nm). In order to characterise the substances causing fouling, size exclusion chromatography is performed for each stirred cell experiment on the feed water, permeate, and concentrate or retentate (see Section 3.5.2 for details on the analysis of the resulting chromatograms and the relation between retention time and eluting compounds/groups of substances).

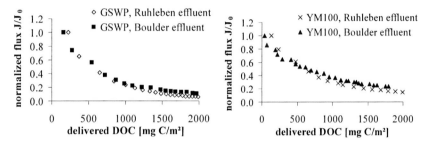

**Figure 4.1**    Flux decline curves for Boulder EfOM and Ruhleben EfOM with a hydrophilic MF membrane (GSWP, left) and a hydrophilic UF membrane (YM100, right)

**Table 4.1**    Fouling indices for Boulder EfOM and Ruhleben EfOM

|  | Boulder EfOM | | Ruhleben EfOM | |
|---|---|---|---|---|
|  | delivered DOC at 25 % flux decline | % of initial flux at 500 mg C/m² | delivered DOC at 25 % flux decline | % of initial flux at 500 mg C/m² |
| YM100 | 292 mg C/m² | 61 | 295 mg C/m² | 59 |
| GSWP | 276 mg C/m² | 58 | 330 mg C/m² | 53 |

Figures 4.2 and 4.3 depict representative organic carbon chromatograms of the feed and permeate samples from the stirred cell tests (in this case with the ultrafiltration membrane YM100) for Boulder EfOM (HPLC-SEC system, Boulder) and Ruhleben EfOM (LC-OCD system, Berlin), respectively. For both effluents, the largest reduction in organic carbon is found for the substances eluting in the PS peak when comparing the feed water sample with the permeate sample collected during the stirred cell experiments. 67 % (Boulder EfOM, Figure 4.2) and 75 % (Ruhleben EfOM, Figure 4.3) of the organic material eluting in this peak are retained by the UF membrane. The reduction in the PS peak is not as pronounced for the MF membrane (50 %) but still more important than for the humic substances and organic acids peaks (Figure 4.4). Hence, large macromolecules such as polysaccharides, possibly proteins, and organic colloids are the most severe foulants in low-pressure membrane filtration of municipal wastewater treatment plant effluents.

It must be noted that only a small percentage of the total DOC delivered to the membrane surface contributes to the fouling layer. Approximately 10 % of the effluent organic matter (measured as TOC) are retained by the micro- and ultrafiltration membranes and thus, are causing the observed loss in permeate flux. This is in the same range as for natural organic matter. Howe & Clark [2002a] report that a maximum of 15 % of the raw water DOC (various surface waters) is retained by MF and UF membranes, concluding that 85-95 % of the DOC must be seen as non-fouling compounds.

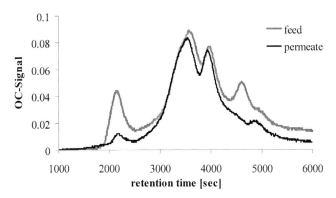

**Figure 4.2**    HPLC-SEC chromatograms of a stirred cell experiment with Boulder EfOM and the ultrafiltration membrane YM100

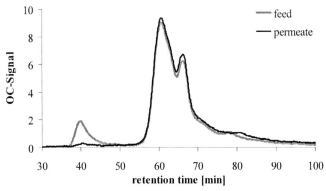

**Figure 4.3**    LC-OCD chromatograms of a stirred cell experiment with Ruhleben EfOM and the ultrafiltration membrane YM100

The TOC concentrations of permeate, feed, and concentrate for the stirred cell experiments using Ruhleben effluent with the hydrophilic MF membrane (GSWP) are, for example, 8.6 mg C/L, 9.5 mg C/L, and 10.2 mg C/L, respectively. These values are obtained with a

TOC analyzer that works with the combustion method (highTOC, Elementar, Germany, see Section 3.6).

When analysing the corresponding LC-OCD chromatograms for this stirred cell experiment (Figure 4.4) and the organic carbon concentrations calculated from the peak areas (Table 4.2), it becomes clear that the difference in TOC content is primarily due to the PS peak. Half of the PS peak content is retained by the membrane, i.e. 0.4 mg C/L of the original 0.8 mg C/L (feed) are measured in the permeate sample, while nearly all the humic substances and organic acids pass the membrane (permeate: 6.6 mg C/L; feed: 6.8 mg C/L). This is supported by the results of the concentrate sample exhibiting a significantly higher PS peak (1.4 mg C/L) and approximately the same amount of humic substances and organic acids (6.9 mg C/L). Based on a mass balance the PS peak in the concentrate sample should amount to 2.4 mg C/L. Hence, it must be assumed that a large amount of substances eluting in the PS peak (1 mg C/L) is contributing to the fouling layer on the membrane.

**Ruhleben effluent with hydrophilic MF (GSWP)**

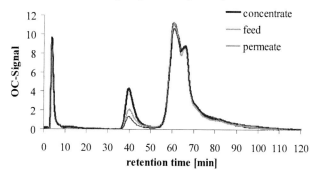

**Figure 4.4**     LC-OCD chromatograms including the by-pass peak for a stirred cell experiment with Ruhleben effluent and the hydrophilic MF membrane (GSWP)

Furthermore, the measurement of the by-pass peak before the injection of the sample into the SEC column allows for the assessment of the so-called hydrophobic organic carbon (HOC) content. The HOC is the part of the organic carbon that adsorbs onto the SEC resin and therefore, is not accessible to the chromatographic analysis. Lipids are an example of such hydrophobic organic substances that can not be separated with size exclusion chromatography. The HOC is calculated by subtracting the CDOC[2] from the by-pass TOC. The by-pass TOC is calculated from the by-pass peak (2-10 min retention time) while the CDOC represents the cumulated organic carbon of all peaks between 36-100 min retention time. As can be seen from Table 4.2 the amount of hydrophobic organic carbon is very

___
[2] CDOC = DOC that can be accessed by size exclusion chromatography

low, i.e. 0.1 mg C/L out of approximately 10 mg C/L. Hence, the organic carbon distribution as detected by the LC-OCD is representative of the DOC in the sample and therefore, it can be assumed that any substances responsible for the fouling of low-pressure membranes can be identified using size exclusion chromatography.

**Table 4.2**     Organic carbon concentrations as calculated from the LC-OCD chromatograms of a stirred cell test with Ruhleben effluent and the hydrophilic MF membrane, GSWP (CDOC corresponds to the area under the chromatogram between 36 – 100 min, HOC is the hydrophobic organic carbon that adsorbs onto the SEC resin)

|                                        | permeate | feed | concentrate |
|----------------------------------------|----------|------|-------------|
| by-pass TOC [mg C/L]                   | 9.4      | 9.5  | 10.3        |
| CDOC [mg C/L]                          | 9.3      | 9.5  | 10.2        |
| HOC = TOC – CDOC [mg C/L]              | 0.1      | 0.0  | 0.1         |
| PS peak [mg C/L]                       | 0.4      | 0.8  | 1.4         |
| HS peak + organic acids peak [mg C/L]  | 6.6      | 6.8  | 6.9         |

## 4.2    Sequential filtration experiments with Boulder EfOM

In order to better understand the size and nature of the foulants, sequential filtration experiments are carried out. The idea of sequential filtration is to remove the fouling causing part of the EfOM and to test the fouling potential of the remaining EfOM, present in the original permeate, on a new identical membrane. If the fouling of the first membrane is caused by sieving alone, the foulants must be larger than the membrane pores and thus, are physically retained. Cake formation would be expected as fouling mechanism because of the accumulation of the foulants on the membrane surface. Pore blockage could also occur if the foulants have molecular sizes similar to the pore size of the membrane. In any case, a second identical clean membrane should theoretically not be fouled by the permeate of the first membrane filtration.

If the second membrane is fouled by the permeate of the first membrane filtration, it must be assumed that the fouling causing substances are smaller than the pore size of the respective membrane. This means that the foulants can theoretically pass the membrane but are retained because of membrane-foulant interactions. In this case, besides physical retention due to size/ steric exclusion, primarily chemical interactions would be assumed, e.g. adsorption of the foulant on the membrane surface and/or in the membrane pores. In the latter case, the fouling mechanism is pore constriction. Another possibility would be the formation of a gel layer on the membrane surface as the foulants agglomerate. Hydrophilic membranes are chosen for the sequential filtration experiments in order to avoid any fouling due to hydrophobic-hydrophobic interactions.

In the first set of sequential filtration tests, bulk Boulder EfOM is used with the hydrophilic UF membrane (YM100). A sample of bulk Boulder effluent is filtered through a clean 100,000 D ultrafiltration membrane and the permeate is collected (= stirred cell test with the first membrane). The permeate is then filtered through a new clean 100,000 D ultrafiltration membrane (second membrane). The resulting flux decline curves are depicted in Figure 4.5.

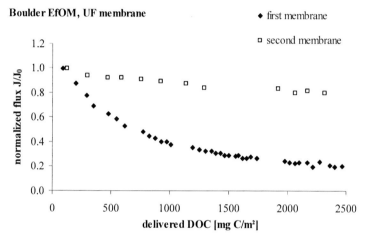

**Figure 4.5**    Flux decline curves of sequential filtration experiments using regenerated cellulose UF membranes (YM100)

The first membrane exhibits a typical flux decline behaviour for the combination of Boulder effluent – UF membrane (see also Figure 4.1 right). For 500 mg C/m² of DOC delivered to the surface of the loose hydrophilic ultrafiltration membrane, the flux declines to about 60 % of its initial flux (Figure 4.5 "first membrane"). The second membrane is fouled to a much lower extent, i.e. 8 % flux decline at 500 mg C/m² delivered DOC. This suggests that most of the foulants are larger than 100,000 D and are retained due to sieving. A linearization of the flux decline data using the characteristic coordinates for cake formation supports this finding (see Chapters 2 and 7 for a more detailed description of the modelling of fouling mechanisms). The characteristic coordinates for cake formation are t/V versus volume. Figures 4.6 and 4.7 depict the plots using these coordinates for the first and second membrane filtration, respectively. A clear linear relationship exists between t/V and V for the first UF membrane filtration as indicated by an $R^2$ value of 0.9996. In contrast, no linearization is obtained with the cake formation model for the second UF membrane filtration nor can the pore constriction model or the pore blockage model explain the data.

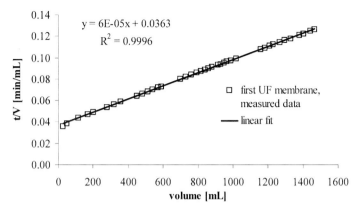

**Figure 4.6**   Linearized flux data for the first UF membrane filtration using characteristics coordinates for cake formation

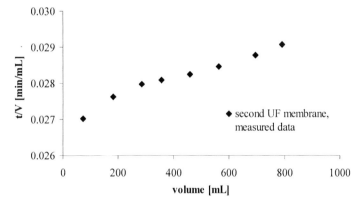

**Figure 4.7**   Linearized flux data for the second UF membrane filtration using characteristics coordinates for cake formation

Depending on the manufacturer, the MWCO of a membrane is determined with various standard molecules (polyethylene glycols, dextrans, etc.). Unfortunately, no exact correlation exists between the molecular weight and the size of a molecule. The molecular size depends primarily on the structure of the molecule, i.e. whether it is more in the shape of a long chain or globular. This results in different ratios of molecular weight to size for the standard substances used for MWCO determination. Hence, no exact correlation exists between the MWCO of a membrane and its pore size. As a rule of thumb, it is often assumed that 100,000 D correspond to approximately 10 nm. Based on diffusion coefficient measurements Howe & Clark [2002b] calculate pore sizes for four membranes of the YM series. Assuming polyethylene glycols are used as standard molecules for the

MWCO determination, they obtain a pore size of 14 nm for the YM100 membrane (MWCO 100,000 D), of 8 nm for the YM30 membrane (MWCO 30,000 D), of 5 nm for the YM10 membrane (MWCO 10,000 D), and of 3 nm for the YM3 membrane (MWCO 3,000 D).

Results of sequential filtration tests with surface waters identify the foulants as compounds of 3-20 nm in size. Substances smaller than approx. 3 nm, i.e. which pass a 3,000 D ultrafiltration membrane (YM3) only contribute to 15 % to the fouling, although they are responsible for 90 % of the DOC. 35 % of the fouling are caused by material retained by a 100,000 D ultrafiltration membrane and therefore, 65 % of the foulants must be in the fraction of 3,000-100,000 D [Howe & Clark 2002b]. In contrast to these findings for surface waters, the fouling causing substances present in the Boulder effluent are at least 10 nm in size. This emphasizes the importance of those EfOM components that can not be attributed to the drinking water NOM such as soluble microbial products (SMP) originating in the biological wastewater treatment process.

In a second set of sequential filtration tests, a membrane with larger pores is used in order to obtain more information on the size of the foulants. A tight hydrophilic MF membrane (MX500) is chosen for these experiments. This membrane has a pore size of 0.05 μm (i.e. 50 nm) and the membrane material is polyacrylonitrile (PAN) which is also very hydrophilic, similar to the regenerated cellulose material of the UF membrane (see Section 3.1). Boulder effluent is filtered through a clean 0.05 μm microfiltration membrane and the permeate is collected (= stirred cell test with the first membrane). The permeate is then filtered through a new clean 0.05 μm PAN microfiltration membrane (second membrane). The resulting flux decline curves are depicted in Figure 4.8.

Stirred cell experiments with bulk effluent organic matter (EfOM) from the WWTP Boulder show significant flux decline (see Section 4.1). This is also the case for the tight hydrophilic microfiltration membrane, MX500. The first membrane of the sequential filtration test is strongly fouled exhibiting a flux decline to approx. 40 % of the initial flux after 500 mg C/m² are delivered to the membrane surface (Figure 4.8). The subsequent filtration of the obtained permeate through a new MX500 membrane also exhibits clear fouling. The flux decline in this second stage of the sequential filtration with the tight microfiltration (MX500) membrane is approx. 20 % at a delivered DOC of 500 mg C/m². Thus, some of the foulants are able to pass a membrane with a pore size of 0.05 μm causing flux decline for a second identical membrane. This suggests that the foulants of Boulder EfOM are in the size of 10-50 nm assuming that the same substances are fouling the MF and UF membranes used (MX500 and YM100, respectively) as is supported by the size exclusion chromatograms. However, a linearization of the flux data reveals cake formation as primary fouling mechanism for the first MF membrane (Figure 4.9), as is the case for the UF membrane too, which points to larger sizes of the fouling causing substances. For the second MF membrane, a linearization of the flux data is achieved with the pore constriction model but also with the cake formation model. This suggests that the fouling causing substances are a heterogeneous mixture with a more or less broad distribution of molecular sizes.

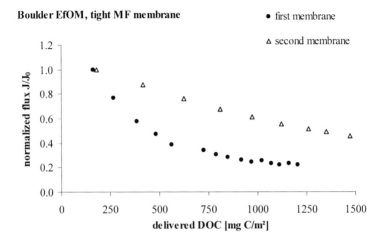

**Figure 4.8**     Flux decline curves for the sequential filtration tests using polyacrylonitrile MF membranes (MX500)

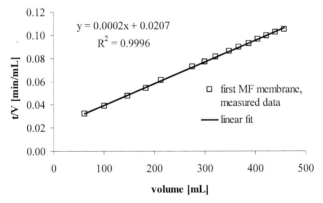

**Figure 4.9**     Linearized flux data for the first MF membrane filtration using characteristics coordinates for cake formation

Additionally, an analysis of the TOC concentrations in the feed, permeate, and retentate samples for the first and second membrane suggests a rejection of larger compounds. The TOC values are given in Table 4.3. Nearly 15 % of the EfOM (as TOC) are rejected during the first membrane filtration with a decrease from 7.4 mg C/L in the feed sample to 6.3 mg

C/L in the permeate sample and a corresponding increase in the retentate sample. In contrast, only 1.5 % of the TOC is retained by the second membrane. A sequential filtration with three or more membranes in series would have been of interest in order to see if the permeate of the second membrane still induces flux decline and fouling with a new identical membrane.

**Table 4.3**   TOC concentrations for feed, permeate, and retentate and fouling indices for the sequential filtration tests with tight MF membranes (MX500)

|  | first membrane | second membrane |
|---|---|---|
| TOC feed [mg C/L] | 7.4* | 6.3** |
| TOC permeate [mg C/L] | 6.3 | 6.2 |
| TOC retentate [mg C/L] | 14.2 | 6.4 |
| delivered DOC @ 25 % flux decline [mg C/m²] | 280 | 625 |
| % of initial flux @ 500 mg C/m² [%] | 45 | 82 |

* bulk Boulder effluent     ** permeate of first membrane filtration

Results of Lee [2003] suggest that an upper limit with regards to the size of foulants from surface waters is found at 450 nm. Stirred cell tests with hydrophilic as well as hydrophobic MF and UF membranes reveal no distinct differences in the flux decline curves for 1 µm and 0.45 µm pre-filtered surface waters. Hence, Lee concludes that dissolved organic matter (DOM: < 0.45 µm) is responsible for the flux loss while particulate organic matter (POM: 0.45 – 1 µm) does not significantly contribute to the membrane fouling.

A sequential filtration of Ruhleben effluent through a 0.45 µm cellulose nitrate filter followed by a 0.1 µm polyether sulfone filter reveals no differences in the organic carbon content of the PS peak. In both filtrate samples the PS peak contains 1 mg C/L. Hence, the foulants eluting in the PS peak must be smaller than 0.1 µm as an identical amount of organic carbon is found after filtration through 0.45 µm and 0.1 µm filters. This means that the foulants present in wastewater treatment plant effluents must be smaller than 100 nm in size assuming a similarity between Ruhleben EfOM and Boulder EfOM, as is suggested by the results of the stirred cell experiments discussed in Section 4.1. Hence, the underlying fouling mechanism in the tight MF (pore size 0.05 µm) and UF (MWCO 100,000 D) membranes is most likely cake formation. In contrast, the two looser MF membranes (pore size 0.22 µm) are more likely to exhibit fouling due to pore constriction and/or a combination of pore constriction and cake formation. An analysis of the underlying fouling mechanisms for all membranes is presented in Chapter 7.

In summary, sequential filtration tests reveal that the fouling causing substances as present in wastewater treatment plant effluents are most likely to have sizes of 10-100 nm. However, it is important to keep in mind that the specific membrane-foulants interactions are likely to vary between different membrane materials. For example, the tight MF membrane (MX500) exhibits a steeper flux decline curve than either the UF membrane

(YM100) or the looser MF membrane (GSWP), although the membrane pore size of the tight MF membrane lies in-between the two latter ones (fouling indices in Tables 4.1 and 4.3). A possible reason could be the different material: the tight MF membrane is made of polyacrylonitrile (PAN) while the other two membranes are made from cellulose.

## 4.3   Hydraulic cleaning of membranes

Often, flux decline cannot be attributed to a single fouling mechanism but is a combination of concentration polarization, deposition onto the membrane surface, and adsorption on or within membrane pores [Yuan and Zydney 2000]. The latter mechanism is mostly irreversible while the first two mechanisms are usually defined as reversible fouling. Generally, a cake layer is more readily removed by backwashing (hydraulic cleaning) while a gel-layer or adsorptive fouling requires chemical cleaning. Hence, besides the flux decline curves, the reversibility of the fouling layer plays an important role in the efficient operation of membrane plants. Ideally, the initial flux is recovered completely after each cleaning cycle. Nevertheless, a complete restoration of the initial flux is usually not possible with backwashing/ hydraulic cleaning alone and a chemical cleaning is eventually needed (see Section 2.1.4).

In order to assess the reversibility of the fouling, stirred cell experiments are carried out. In a hydraulic cleaning experiment with a hydrophilic UF membrane (YM100) fouled by Boulder EfOM the flux has been restored to 117 % of its initial value. Ultra-pure water is used for the hydraulic cleaning. Similar findings are reported by Lee [2003] for the same membrane fouled with natural organic matter (NOM). The author concludes that UF membranes are fouled by easily removable cake/gel layers. However, an examination of the UF membrane after the hydraulic cleaning reveals that the thin membrane layer has been partly detached from its support layer. This is most likely the reason for the higher flux after the hydraulic cleaning and one reason why backwashes in large-scale membrane plants are only done for symmetric hollow-fiber membranes. Besides the loose ultrafiltration membrane (YM100), the other membranes used in this research are symmetric membranes, i.e. no separate thin layer exists and thus, hydraulic cleaning experiments should be able to simulate full-scale backwashes.

Figure 4.10 shows the flux recovery after different hydraulic cleaning procedures for the tight hydrophilic microfiltration membrane (MX500) with bulk Boulder EfOM as feed water. Three different procedures are tested in this case:
   a) halt of permeate production by releasing the pressure,
   b) hydraulic cleaning with permeate at 1 bar TMP, and
   c) hydraulic cleaning with ultra-pure water at 1 bar TMP.

The idea behind procedure a) is that by interrupting the filtration process, foulants accumulated in a cake layer would be able to diffuse back into the bulk solution, thereby, diminishing the filtration resistance associated with the cake layer. To halt the permeate production, the pressure is released from the stirred cell testing set-up. Concomitantly the stirring of the sample is paused and the 200 mL of feed water present in the filtration cell at that time are left unstirred in contact with the membrane surface. If the sample is left in the filtration cell for > 14 h (unstirred and non-pressurized) 75 % of the initial flux can be recovered (Figure 4.10, Point a). This supports the hypothesis of the formation of a cake layer (see also Section 4.2) versus a gel layer as the majority of the fouling is reversible by a simple pressure release. However, after resuming the filtration, the rate of flux decline is higher than beforehand as the foulants are not removed from the system.

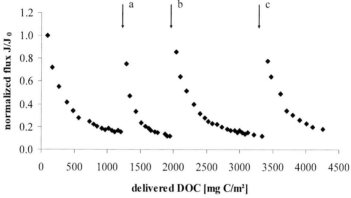

**Figure 4.10**    Recovery of initial flux after different cleaning procedures using Boulder effluent as feed water with the MX500 tight MF membrane: a) after leaving the sample unstirred and non-pressurized in the filtration cell for > 14 h, b) after hydraulic cleaning with permeate, c) after hydraulic cleaning with ultra-pure water

Secondly, hydraulic cleanings are performed (Points b and c). These simulate a backwash in full-scale membrane plants. Two different procedures have been investigated: a hydraulic cleaning with permeate as is usually done in full-scale plants (Point b) and a hydraulic cleaning with ultra-pure water (Point c). The hydraulic cleaning with permeate results in a flux recovery of 86 % while the hydraulic cleaning with ultra-pure water restores the flux to 78 % of its initial value. Thus, both hydraulic cleaning schemes do not restore the initial flux completely, although, the flux recovery is higher than with pressure release alone. This is in accordance to the experience in full-scale plants where periodic backwashes are used to enhance the membrane performance but more thorough chemical cleanings are periodically needed to restore the initial flux (see Section 2.1.4).

Nevertheless, the hydraulic cleaning with ultra-pure water has been expected to restore the initial flux completely or at least to an higher extent than a hydraulic cleaning with permeate. Calcium ($Ca^{2+}$) often plays an important role in the fouling of membranes [Yuan and Zydney 2000] as it serves as bridge between the membrane surface and organic foulants. Since cleaning with ultra-pure water would wash out any calcium, the fouling layer should be completely removed if bridging is of importance. In contrast, a hydraulic cleaning with permeate would not change the calcium concentration because the concentration of these ions and salt ions in general is not changed during low-pressure membrane filtration (see Chapter 2). However, the hydraulic cleaning with ultra-pure water does not restore the flux to its initial value suggesting that changes in intermolecular electrostatic interaction due to the presence/absence of $Ca^{2+}$ are not of importance here.

This is supported by a stirred cell experiment using the transphilic acids (TPI-A) fraction of Boulder EfOM (for details on this fraction see Section 5.1) with and without the addition of calcium. The sample solutions are obtained by dissolving 10 mg/L of the TPI-A isolate in ultra-pure water yielding a DOC of approximately 4.5 mg C/L. The pH is adjusted with NaOH to 6.5 and the conductivity is raised to 750 μS/cm with sodium sulphate in a phosphate buffer matching the conductivity of Boulder effluent. For the addition of calcium, $CaCl_2$ is used and the concentration is adjusted to 1mM $Ca^{2+}$. The same membrane as in the hydraulic cleaning experiment is used, i.e. the tight hydrophilic MF membrane (MX500). Figure 4.11 depicts the resulting flux decline curves. No difference is detectable between the two curves and therefore, it is concluded that calcium has no influence on the fouling behaviour. In accordance to these findings, Lee [2003] reports that an influence of $Ca^{2+}$ is not detected: a hydrophobic (polyether sulfone) and a hydrophilic (regenerated cellulose) UF membrane are fouled with either the hydrophobic NOM fraction (isolated by XAD-8) of a surface water or the protein albumin. Addition of 3 mM $Ca^{2+}$ as $CaCl_2$ did neither enhance nor reduce the amount of flux decline and thus fouling.

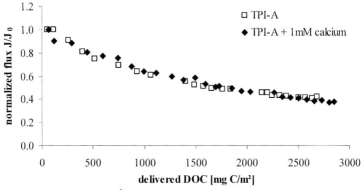

**Figure 4.11**    Influence of $Ca^{2+}$ on membrane fouling for the tight MF membrane (MX500) with the transphilic acids fraction of Boulder EfOM

Size exclusion chromatograms of the backwash water (i.e. the permeate and ultra-pure water used for the hydraulic cleanings) are of special interest for characterizing the fouling layer. Figure 4.12 depicts the chromatograms obtained from hydraulic cleaning with ultra-pure water (left) and with permeate (right) for the stirred cell experiment with the tight microfiltration membrane described above. The upper curves represent the organic carbon response corresponding to the left axis while the lower curves show the UV absorbance at 254 nm (right axis). In both cases, a clear peak in the PS region (~ 1900 sec retention time) is detected in the organic carbon response. This peak clearly dominates the chromatogram of the ultra-pure backwash water (Figure 4.12 left) while only a small amount of humic substances (HS) are detected. The sample of the hydraulic cleaning with permeate, on the other hand, exhibits a clear HS peak, too. Because these substances are not retained by the membrane during membrane filtration, they are present in the permeate and thus, also appear in the backwash sample.

Additionally, a clear UV absorption peak is detected in the polysaccharides peak area (Figure 4.12, lower curves). This is an indication for double bonds and structures with delocalized $\pi$-electrons. However, polysaccharides do not have the structures for UV absorbance at 254 nm and therefore, other large macromolecules must elute in this peak besides polysaccharides. This could be either large proteins or organic colloids. Effluent organic matter contains organic colloids, for example, in the form of bacterial peptidoglycan residues, i.e., cell wall fragments that are made of polysaccharides and proteins [Schlegel 1992]. NOM colloids have been reported to be severe foulants for ultrafiltration membranes [Habarou et al. 2001]. Thus, the chromatograms of the hydraulic cleaning tests support the hypothesis that polysaccharides and larger macromolecules are the most severe foulants of low-pressure membranes.

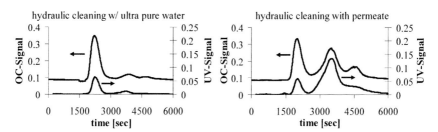

**Figure 4.12**    HPLC-SEC chromatograms of the permeate used for the first hydraulic cleaning test (right) and of the ultra-pure water used in the second hydraulic cleaning (left); upper curves show OC-signal and lower curves UV-signal

# Chapter 5     Effluent organic matter (EfOM): isolates and their fouling properties and character

Using the isolation procedure described in Section 3.4, three different fractions of the bulk Boulder effluent are isolated: colloids, hydrophobic acids (HPO-A) and transphilic acids (TPI-A). In this chapter, first a description of the characteristics of these fractions is given (Section 5.1) followed by results and discussion of the stirred cell experiments (Section 5.2). Some of the data in this chapter has been presented at various conferences [Reichenbach et al. 2001, Laabs et al. 2002, Laabs et al. 2003].

## 5.1    Isolates and their characteristics

For the two fractions, hydrophobic acids (HPO-A) and transphilic acids (TPI-A), elemental analysis[3] and [13]C-NMR analyses[4] have been carried out. Unfortunately, the colloidal isolate did not amount to enough material to allow for these analyses. Tables 5.1 and 5.2 give the results of the [13]C-NMR analyses and elemental analyses, respectively. The [13]C-NMR spectra as well as the elemental analysis show similarities and differences between hydrophobic acids and transphilic acids. Both isolates exhibit a similar percentage (65 % and 66 %, respectively) of carbon in the range between zero ppm and 90 ppm. The difference, however, is that nearly 50 % of the organic carbon for the HPO-A fraction is aliphatic carbon (0-62 ppm) and only 17 % accounts for heteroaliphatic carbon (62-90 ppm) while the two peaks have similar areas for the TPI-A isolate, 34 % and 32 % respectively. Thus, the transphilic acids contain more hetero atoms such as oxygen, nitrogen, and sulphur. This is supported by the elemental analysis with higher oxygen, nitrogen, and sulphur contents in the transphilic acids than in the hydrophobic acids (see Table 5.2). Carbon and oxygen content are approximately equivalent (~ 44 %) for the transphilic acids. Consequently, carboxylic (160-190 ppm) and anomeric (90-110) carbon are higher in the transphilic isolate, too.

A higher carbon content for the hydrophobic acids (HPO-A), 52 % compared to 45 % in the transphilic acids (TPI-A), is consistent with the more aromatic character of this fraction: The peak area from 110-160 ppm (aromatic carbon) accounts for 18 % of the carbon in the HPO-A isolate whereas the TPI-A isolate has only 14 % aromatic carbon. For both isolates only small amounts of ketonic carbon are detected.

---

[3] Analyses are done by Huffman laboratories Inc., Golden (Colorado, USA)

[4] Analyses are done at the Denver Federal Center (Colorado, USA) of the United States Geological Survey (USGS)

**Table 5.1**                    $^{13}$C-NMR data in percent (%) carbon with 1 msec contact time

|                                                          | HPO-A | TPI-A |
| --- | --- | --- |
| 0 –   62 ppm (aliphatic C)                               | 48.3  | 34.2  |
| 62 –   90 ppm (heteroaliphatic C, e.g. alcohols)         | 16.8  | 32.0  |
| 90 – 110 ppm (anomeric C)                                | 5.4   | 7.0   |
| 110 – 160 ppm (aromatic C)                               | 17.8  | 13.8  |
| 160 – 190 ppm (carboxylic C)                             | 9.4   | 11.2  |
| 190 – 220 ppm (ketonic C)                                | 2.2   | 1.7   |

**Table 5.2**        Elemental Analysis of the hydrophobic acids (HPO-A) and transphilic acids
(TPI-A) (ash free content shown in %)

|           | HPO-A | TPI-A |
| --- | --- | --- |
| Carbon    | 52.1  | 44.6  |
| Hydrogen  | 5.79  | 5.23  |
| Oxygen    | 36.6  | 43.5  |
| Nitrogen  | 2.69  | 3.37  |
| Sulphur   | 2.86  | 3.33  |
| H/C ratio | 1.33  | 1.41  |
| O/C ratio | 0.53  | 0.73  |
| N/C ratio | 0.044 | 0.065 |

Thus, the $^{13}$C-NMR and elemental analyses confirm the intended results of the isolation procedure. The HPO-A isolate exhibits a more hydrophobic character with a high content of aromatic structures. The XAD-8 resin which is used to isolate the hydrophobic acid fraction, is known for its effective adsorption capacity for humic substances. In the case of aquatic humic substances this means primarily fulvic acids as humic acids amount to only approximately 10 % of the dissolved humic substances in water [Aiken 1985]. In comparison, the TPI-A isolate which is isolated with the XAD-4 resin exhibits generally a less hydrophobic character with more hetero atoms. This is in accordance with the findings of Debroux [1998] who reports for three different wastewater effluents that the XAD-8 isolate has higher SUVA values and aromaticity but smaller amounts of nitrogen, sulfur, and carboxylic carbon than the corresponding XAD-4 isolates. In the case of Boulder effluent the SUVA values are 2.19 L mg$^{-1}$m$^{-1}$ for the hydrophobic acids fraction (HPO-A) and 1.46 L mg$^{-1}$m$^{-1}$ for the transphilic acids fraction (TPI-A).

Interestingly, Debroux [1998] reports higher molecular weights for the XAD-4 isolate than for the XAD-8 isolate using size exclusion chromatography. This appears to be also the case here with the peak maxima of the transphilic acids eluting somewhat earlier than those of the hydrophobic acids (Figure 5.1). However, according to Aiken et al. [1992] the XAD-8 isolate contains mainly humic and fulvic acids, while the XAD-4 isolate consists of slightly smaller and less hydrophobic compounds. This apparent contradiction could be due to the fact that the molecules elute according to their size including any hydrating water molecules and not according to their molecular weight in size exclusion chromatography. Additionally, it must be kept in mind that any molecular weight calibration in size exclusion chromatography depends on the standards used. It is especially difficult to find adequate calibration standards for humic substances as the structure of these compounds is not completely identified, yet. As can be seen from Table 3.6, calibrations with dextrans, polyethylene glycols (PEG) or globular proteins yield differing separation ranges for the same size exclusion column.

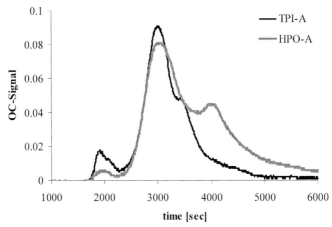

**Figure 5.1**    Organic carbon chromatograms for HPO-A and TPI-A isolates (measured with the HPLC-SEC system, Boulder)

However, for both isolates the so-called humic substances (HS) peak dominates the chromatogram (Figure 5.1). This supports the assumption of Aiken et al. [1992] that humic-like substances are adsorbed onto the XAD-4 resin. Nevertheless, a small PS peak appears in both isolates although such large molecules would be expected to be found in the hydrophilic fraction (see Section 3.4) which has not been studied in this research due to its high salt content. The large macromolecules of the PS peak rather, in this research, have been isolated in the colloidal fraction.

Because of the presence of humic-like material in the isolated colloids using the standard dialysis bag with a molecular weight cut-off (MWCO) of 3500 Dalton (D), two larger dialysis bags with 6-8 kD and 12-14 kD MWCOs have been used. Colloids isolated with these two dialyis bags are referred to as "colloids > 6-8 kD" and "colloids > 12-14 kD", respectively (see Section 3.4 for the isolation procedure).

Size exclusion chromatography, however, reveals no real differences between these two isolates. Figure 5.2 depicts the LC-OCD organic carbon chromatograms for both isolates.

The main peak elutes close to the void volume in the area of the so-called PS peak. A second, less important peak elutes after 58 min retention time (peak maximum). The HW-55S size exclusion column is used in the LC-OCD (see Section 3.5). Based on a calibration with polyethylene glycols (PEGs), the compounds eluting in the peak after 58 min retention time have an approximate molecular weight of 8000 D. If the calibration is done with dextrans, a peak after 58 min retention time would correspond to a molecular weight of approximately 12,500 D. This correspond quite well to the stated MWCO of the two dialysis bags with 6-8 kD and 12-14 kD. Knowing that molecular weight cut-offs are never 100 % precise and depend very strongly on the specific substances, it is assumed that the two colloids isolates can be regarded as equal to one another, especially because this smaller molecular weight peak is less important in overall DOC than the peak eluting after 41 min retention time.

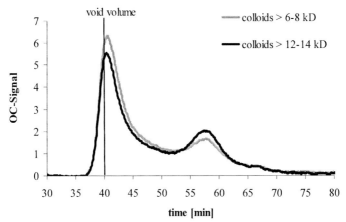

**Figure 5.2**    LC-OCD chromatograms of the two colloidal isolates: colloids > 6-8 kD and colloids >12-14 kD

The substances eluting in the first peak (after 41 min retention time) are at least an order of magnitude larger with a molecular weight of > 182,000 D with a calibration using PEGs (peak maximum of largest standard: PEG 182,000 D at 43 min retention time) or > 124,000 D with a calibration using dextrans (peak maximum of largest standard: dextran 123,600 D at 47 min retention time). However, the void volume of the column has been determined to be at 40 min retention time using blue dextran with an approximate molecular weight of 2,000,000 D. Since the first peak of the colloids isolates elutes close to the void volume of the column, it is likely that the corresponding substances are much larger than the separation range of the column. Nevertheless, they are unlikely to be larger than 1-2 µm, otherwise they would have been retained by the in-line pre-filter of the LC-OCD system (see Section 3.5). A size exclusion column with a larger pore size, and thus a wider separation range, should be better able to further differentiate the colloidal peak into organic colloids and dissolved polysaccharides (and proteins). On the pre-condition that large enough calibration standards are available, this would also allow for a more precise determination of the molecular weight/ size of the colloidal compounds.

Furthermore, attenuated total reflectance Fourier transform infra-red spectroscopy (ATR-FTIR) has been performed on all three fractions, i.e. colloids, hydrophobic and transphilic acids. The resulting spectra are depicted in Figure 5.3. The –OH (alcohols) stretching vibration at 3440 $cm^{-1}$ and the –CH (alkanes) stretching vibration at 2950 $cm^{-1}$ can be found in all three organic isolates. In the fingerprint region, the main peak of HPO-A and TPI-A is at 1740 $cm^{-1}$ which is typical for carboxylic (–CO-OH) and ketonic groups as found in humic substances. A shoulder to the right of the peak at 1650 $cm^{-1}$, is due to either aromatic structures (C=C) or amides (–CO-N). However, if amides are present another peak at 1540 $cm^{-1}$ would be detected. This is only the case for the organic colloidal material. The characteristic fingerprints of the organic colloidal material are the two peaks at 1650 $cm^{-1}$ and 1080 $cm^{-1}$. As a smaller peak at 1540 $cm^{-1}$ is detected, the peak at 1650 $cm^{-1}$ can be attributed to amide groups, i.e. proteins [Jarusutthirak 2002]. The peak at 1080 $cm^{-1}$ stems from C-O-C bonds (ether) in the molecules as present in polysaccharides.

Besides the organic colloids, inorganic colloids might remain in the dialysis bag. A presence of silica based inorganic colloids would result in a strong silica response around 400 $cm^{-1}$ [Leenheer 2002]. Mineral clay colloids containing $SiO_2$ also exhibit a peak at 1040 $cm^{-1}$, however no peak at 2950 $cm^{-1}$ should be present if this peak is due to inorganic colloids [Jarusutthirak 2002]. As this is not the case, it is assumed that the isolated colloids consist of organic material with polysaccharides and proteins as components.

Thus, due to the more hydrophobic character of the HPO-A fraction, a higher fouling potential of this isolate is expected on hydrophobic membranes because of hydrophobic-hydrophobic interactions. The colloids isolates would probably cause the highest flux decline of the three fractions investigated from Boulder effluent since they consist only of compounds eluting in the PS peak (see Section 3.4) and show characteristic fingerprints of proteins and polysaccharides in ATR-FTIR (Figure 5.3). As seen in the previous chapter the substances of the PS peak are responsible for most of the fouling behaviour in the bulk effluent samples.

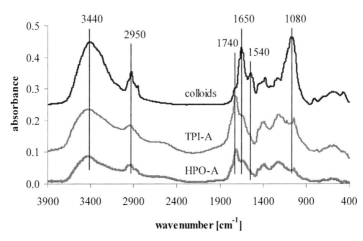

**Figure 5.3**     Fourier transformed infra-red spectra of freeze-dried isolates from Boulder wastewater treatment plant effluent: colloids (top), transphilic acids (middle), hydrophobic acids (bottom)

## *5.2     Stirred cell tests with effluent isolates*

With the knowledge from characterisation of the isolates, stirred cell experiments are performed to assess the fouling potential of the isolates. Wastewater treatment plant (WWTP) bulk effluent organic matter (EfOM) has a significant fouling potential with low-pressure membranes, i.e. micro- and ultrafiltration membranes, as seen in the previous chapter (Chapter 4). Figure 5.4 depicts representatively the flux decline curve for Boulder effluent (Colorado, USA) and its two fractions, HPO-A (hydrophobic acids) and TPI-A (transphilic acids), with a hydrophilic ultrafiltration membrane (YM100) made of regenerated cellulose (see Table 5.3 for the pH, conductivity, DOC content, and SUVA of the feed waters for each of the stirred cell tests). For the bulk Boulder effluent the initial flux $J_0$ (500 L h$^{-1}$m$^{-2}$) declines to 60 % after a delivered DOC[5] of 500 mg C/m². In comparison, the two isolates exhibit less step flux decline curves than the bulk Boulder EfOM. This is expected since the two isolates represent only sub-sets of the bulk EfOM. The transphilic acids fraction has a higher fouling potential on the UF membrane than the hydrophobic acids fraction which might be related to the higher nitrogen content of the TPI-A isolate. After a delivered DOC of 500 mg C/m² the flux decline amounts to 25 % for the TPI-A fraction and to 15 % for the HPO-A fraction. Yuan and Zydney [2000] report similar findings for filtration of humic acid solutions with ultrafiltration membranes. Using different prefiltration membranes they conclude that large humic acid aggregates contribute only to a small extent to the fouling of ultrafiltration membranes.

---

[5] delivered DOC [mg C/m²] = DOC$_{feed}$ [mg/L] * permeate volume [L] / membrane surface area [m²]

**Table 5.3**    Feed water pH, conductivity, DOC content, and SUVA for stirred cell tests
with Boulder EfOM, HPO-A, and TPI-A (YM100 membrane)

|  | pH | conductivity [$\mu$S/cm] | DOC [mg/L] | SUVA [L/mg m] |
|---|---|---|---|---|
| Boulder EfOM | 7.2 | 636 | 4.9 | 2.32 |
| HPO-A | 5.9 | 8.5 | 4.9 | 2.16 |
| TPI-A | 5.7 | 10.3 | 4.1 | 1.55 |

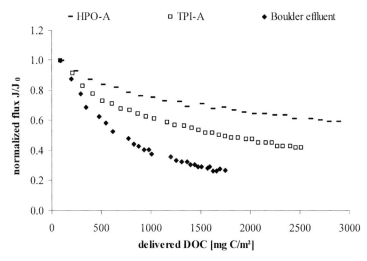

**Figure 5.4**    Flux decline curves for Boulder effluent and its hydrophobic acids fraction
(HPO-A) and transphilic acids fraction (TPI-A) with a hydrophilic UF
membrane (YM100)

With both Boulder effluent fractions (HPO-A and TPI-A) stirred cell experiments are
performed using all four membranes: the hydrophilic ultrafiltration membrane (YM100),
the tight hydrophilic microfiltration membrane (MX500), the loose hydrophilic
microfiltration membrane (GSWP), and the loose hydrophobic microfiltration membrane
(GVHP). Figures 5.5 and 5.6 show the flux decline curves with the four membranes for the
transphilic and hydrophobic acids isolates, respectively.

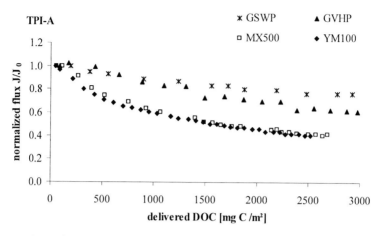

**Figure 5.5**    Flux decline curves for the transphilic acids fraction (TPI-A) with four different membranes: GSWP (hydrophilic, MF), GVHP (hydrophobic, MF), MX500 (hydrophilic, tight MF), and YM100 (hydrophilic, UF)

The effluent transphilic acids have a stronger fouling potential on the ultrafiltration and tight microfiltration membranes than with the loose microfiltration membranes (Figure 5.5). This is true regardless of the membrane material, i.e. regenerated cellulose (YM100) and polyacrylonitrile (MX500) exhibit virtually identical flux decline curves. Additionally, a significant difference in hydrophobicity appears to play only a minor role in the fouling behaviour of the transphilic acids fraction. The two flux decline curves of the loose MF membranes are nearly identical up to a delivered DOC of 1300 mg C/m² although the GSWP membrane is very hydrophilic with a contact angle of 19° while the GVHP membrane has a very hydrophobic surface exhibiting a contact angle of 83°.

The situation is somewhat different with the effluent hydrophobic acids fraction (Figure 5.6). All four flux decline curves are much more similar than for the transphilic acids isolate. The fouling behaviours with the loose ultrafiltration (YM100) and tight microfiltration (MX500) membranes are similar again but with a slightly higher fouling of the MF membrane (MX500) at higher delivered DOC values. In contrast to the TPI-A fraction, the two loose microfiltration membranes are not fouled to the same extent. The GSWP (hydrophilic MF) membrane has the lowest fouling with the hydrophobic acids fraction, as with the transphilic acids fraction. However, the most severe membrane fouling by the HPO-A isolate occurs with the hydrophobic MF membrane (GVHP). This supports the hypothesis that the hydrophobic fraction of wastewater effluent fouls hydrophobic membranes due to hydrophobic-hydrophobic interactions.

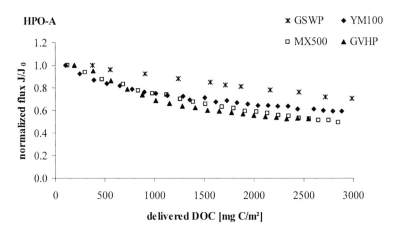

**Figure 5.6**     Flux decline curves for the hydrophobic acids fraction (HPO-A) with four different membranes: GSWP (hydrophilic, MF), GVHP (hydrophobic, MF), MX500 (hydrophilic, tight MF), and YM100 (hydrophilic, UF)

**Table 5.4**     Fouling data for hydrophobic acids (HPO-A) and transphilic acids (TPI-A) isolates with the four membranes YM100 (hydrophilic, UF), MX500 (hydrophilic, tight MF), GSWP (hydrophilic, loose MF), and GVHP (hydrophobic, loose MF); two fouling indices are given: i) the amount of delivered DOC to the membrane surface at 25 % flux decline, ii) the percentage of the initial flux ($=J/J_0*100$) after 500 mg C/m² are delivered to the membrane surface

|  | HPO-A | | TPI-A | |
| --- | --- | --- | --- | --- |
|  | delivered DOC at 25 % flux decline [mg C/m²] | % of initial flux at delivered DOC of 500 mg C/m² [%] | delivered DOC at 25 % flux decline [mg C/m²] | % of initial flux at delivered DOC of 500 mg C/m² [%] |
| YM100 | 1015 | 85 | 470 | 74 |
| MX500 | 980 | 86 | 520 | 76 |
| GSWP | 2450 | 97 | 2930 | 99 |
| GVHP | 875 | 89 | 1495 | 95 |

Besides the clear difference in the fouling behaviour with the hydrophobic MF membrane, a comparison of the fouling data for both Boulder effluent fractions is presented in Table 5.4. Two parameters are extracted from the flux decline curves: i) the amount of delivered DOC at 25 % flux decline and ii) the percentage of the initial flux ($=J/J_0*100$) after 500 mg C/m² are delivered to the membrane surface. A higher amount of delivered DOC at 25 % flux decline means that less fouling occurred as does a higher value for the percentage of initial flux after 500 mg C/m² of delivered DOC.

These comparisons give similar results for the loose ultrafiltration (YM100) and tight microfiltration (MX500) membranes, with the transphilic acids fraction (TPI-A) having a higher fouling potential than the hydrophobic acids fraction (HPO-A). In contrast, the loose MF membranes (GSWP and GVHP) are more severely fouled by the HPO-A isolate. This is in accordance with the findings of Yuan and Zydney [2000]. They state that large humic acid aggregates contribute only to a small extent of the fouling of ultrafiltration membranes while they are responsible for a large amount of the fouling of microfiltration membranes [Yuan and Zydney 1999]. However, in this research, only the hydrophobic GVHP membrane exhibits a similar difference between the two effluent fractions as that seen with the tighter membranes. The loose hydrophilic MF membrane (GSWP) is fouled by both fractions to nearly the same low extent with a flux decline of merely 1-3 % after a delivered DOC of 500 mg C/m².

In contrast to these results, Habarou et al. [2001] found for natural organic matter (NOM) isolates with the YM100 ultrafiltration membrane the least fouling for the XAD-4 isolate. Thus, concerning the fouling characteristics of EfOM, not only NOM is of interest, but other compounds play a role too. These compounds have been identified as synthetic organic compounds (SOC) and disinfection by-products (DBP), which are added by the consumer and during the water treatment process, as well as soluble microbial products (SMP), which are produced in the wastewater treatment plant itself [Drewes and Fox 1999]. Especially the soluble microbial products present in WWTP effluents distinguish EfOM from organic matter of surface waters.

The reasons for the differences in the fouling potential of the transphilic acids fraction and the hydrophobic acids fraction lie, on one hand, in the higher percentage of oxygen, nitrogen, and sulphur for the TPI-A isolate and the higher hydrophobicity of the HPO-A isolate as shown by the elemental analysis and $^{13}$C-NMR spectroscopy. On the other hand, the molecular weight cut-off or pore size of the membranes determines, to a large extent, the fouling and the prevailing fouling mechanisms. The analysis of the fouling mechanisms is the subject of Chapter 7.

Size exclusion chromatography with online UV and DOC detection of the bulk effluent organic matter (EfOM) and the two fractions reveals that compounds eluting in the so-called PS peak (including organic colloids, polysaccharides and large proteins) play the most important role in fouling low-pressure membranes. Figure 5.7 depicts, for example, the organic carbon chromatograms of a stirred cell experiment with the TPI-A fraction and the hydrophilic ultrafiltration membrane (YM100). When comparing the feed solution with the permeate after 20 min of filtration, the two chromatograms look very similar. Nevertheless, the retentate sample (100 mL of sample solution that remain in the stirred cell at the end of the filtration experiment) shows that primarily compounds eluting in the PS peak (peak maximum after approx. 1900 sec retention time) are retained by the membrane. Thus, organic colloids and polysaccharides can be seen as major foulants in low-pressure membrane filtration since only compounds that are rejected by the membrane can contribute to its fouling. Hence, the isolated colloids should contribute the most to the fouling of low-pressure membranes by wastewater treatment plant effluents as this fraction exhibits primarily a large peak in the PS range (elution area of polysaccharides, organic colloids, and proteins) in the chromatograms (see Figure 3.4).

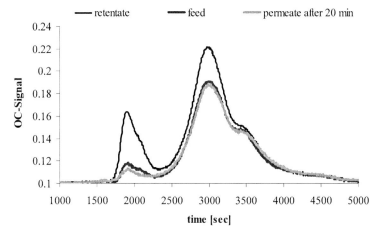

**Figure 5.7**    Size exclusion chromatograms for the stirred cell experiment using 10 mg/L TPI-A isolate with an ultrafiltration membrane (YM100): feed = solution with 10 mg TPI-A/ L ultra-pure water; permeate after 20 min = permeate of stirred cell test collected after 20 min filtration time; retentate = 100 mL remaining in stirred cell at end of filtration (time = 100 min; filtered volume = 1750 mL)

Stirred cell experiments with the colloids are performed with either the "colloids > 6-8 kD" isolate or with the "colloids > 12-14 kD" isolate (see Section 3.4 for the isolation procedure); no significant differences between the two isolates have been detected with the LC-OCD (see Section 5.1). In order to ensure that no other substances interfere with the membrane material, 10 mg of lyophilized colloids are re-dissolved in 1 L ultra-pure water resulting in an organic carbon content of ~ 2 mg/L. The pH is adjusted to 6.6 ± 0.3 and the conductivity to 715 μS/cm ± 15 μS/cm to match the conditions of the bulk Boulder effluent. Although the colloid solutions have been stirred for 24 hours and sonicated for 20 minutes, not all the lyophilized material re-dissolved completely. Hence, the solutions are divided into two portions and stirred cell tests are carried out with unfiltered and pre-filtered colloids solutions (pre-filtration consists of filtration through 1 μm glass fiber filters for the 'colloids > 6-8 kD' sample and through 0.45 μm cellulose nitrate filters for the 'colloids > 12-14 kD' sample to remove non-dissolved particles). The difficulty of re-solubilising the freeze-dried colloids is, according to Buffle and Leppard [1995b], probably due to structural changes induced by the lyophilization of the isolated colloids. However, according to Rauch [2003], problems with re-solubilization may be controlled by acidifying the sample during dialysis. Another possibility would be to follow a more complex isolation procedure as proposed by Leenheer et al. [2000]. Neither has been done for this research because the re-solubilization problems occurred only after the isolation procedure had been finished.

A difference between a pre-filtration through 1 µm glass fiber filters and a pre-filtration through 0.45 cellulose nitrate filters is not observed for the colloids. Table 5.5 presents the SUVA values for the two colloid isolates (colloids > 6-8 kD and colloids > 12-14 kD) after re-solubilization (unfiltered samples) and additional pre-filtration (filtered samples) to remove any remaining particles, i.e. lyophilized colloids that did not re-dissolve. In both cases the SUVA values for the pre-filtered samples are very low with 0.57 L mg$^{-1}$m$^{-1}$ (colloids > 6-8 kD) and 0.56 L mg$^{-1}$m$^{-1}$ (colloids >12-14 kD), respectively. This supports the LC-OCD results that nearly no aromatic structures such as humic-like material are part of the colloidal isolates (see Section 5.1). The SUVA values of the unfiltered samples are approximately three times those of the filtered samples. However, this is most likely due to the interference of particles during the UV measurements (light scattering).

**Table 5.5**    SUVA values before and after the pre-filtration step for the two isolates: colloids > 6-8 kD and colloids > 12-14 kD

|                    | Colloids > 6-8 kD | | Colloids > 12-14 kD | |
|--------------------|------------|-------------------|------------|---------------------|
|                    | unfiltered | 1 µm filtered     | unfiltered | 0.45 µm filtered    |
| SUVA [L/(mg m)]    | 1.59       | 0.57              | 1.76       | 0.56                |
| TOC [mg/L]         | 2.2        | 1.6               | 1.9        | 1.2                 |

The results of the stirred cell experiments using unfiltered and pre-filtered colloids with either a hydrophobic (GVHP) or hydrophilic (GSWP) loose microfiltration membrane are depicted in Figure 5.8. For both membranes, the unfiltered colloids samples, i.e. with undissolved particles in them, exhibit a much steeper flux decline than the pre-filtered samples. The significant difference between the pre-filtered and unfiltered colloids samples can be solely attributed to the particles in the latter ones. The pre-filtration step removes approximately 30 % of the organic carbon content. However, the difference in organic carbon concentration is accounted for in Figure 5.8 as the flux decline curves are plotted as a function of the DOC that is delivered to the membrane surface. Thus, the particles themselves must contribute to a large extent to the fouling of the membranes. Their fouling potential is more or less independent of the membranes' hydrophobicity as the flux declines very quickly to 10 % of the initial flux with both microfiltration membranes.

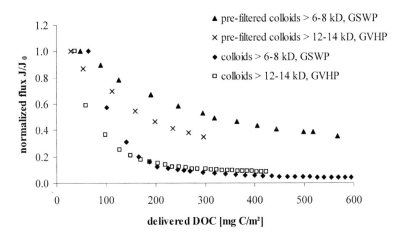

**Figure 5.8**    Flux decline curves for colloids isolates with the hydrophobic (GVHP)
and hydrophilic (GSWP) MF membranes; the pre-filtered samples are
filtered after re-solubilization prior to the stirred cell test to remove
remaining particles

On the other hand, the pre-filtered colloids samples do show a difference in their fouling
behaviour with the two loose microfiltration membranes. A flux decline of 25 % is
achieved after a delivered DOC of only 70 mg $C/m^2$ with the hydrophobic membrane
(GVHP) while 127 mg $C/m^2$ are necessary with the hydrophilic membrane (GSWP). This
is a very small amount of organic carbon to induce such a loss in permeate flux compared
to the effluent acids fractions HPO-A and TPI-A where between nearly 500-3000 mg $C/m^2$
are needed for the same flux reduction (see Table 5.6 compared to Table 5.4). Due to this
extremely high fouling potential of the colloids isolate, the second fouling index (% of
initial flux) in Table 5.5 is taken after 250 mg $C/m^2$ delivered DOC instead of 500 mg $C/m^2$
delivered DOC as with the effluent acids fractions. In accordance to these results, Habarou
et al. [2001] report NOM colloids as the most severe foulants on ultrafiltration membranes.
Interestingly, the colloids isolates foul the membranes to an even higher extent than the
bulk effluents (see Chapter 4). This supports the hypothesis, that the main foulants of low-
pressure membranes are compounds eluting in the PS peak. These compounds are large
polymers such as those present in the EPS (extracellular polymeric substances) produced
by the microorganisms of activated sludge. A more detailed investigation on the influence
of the EPS on low-pressure membrane fouling has been done on two membrane bio-reactor
(MBR) pilot plants, as MBRs usually have higher biomass concentrations (see Chapter 6).

**Table 5.6**    Fouling indices for colloids isolates: i) amount of DOC delivered to the membrane surface at 25 % flux decline ($J/J_0=0.75$) and ii) remaining percentage of initial flux ($=J/J_0*100$) after a delivered DOC of 250 mg C/m²

|  | colloids > 6-8 kD GSWP membrane | | colloids > 12-14 kD GVHP membrane | |
| --- | --- | --- | --- | --- |
|  | unfiltered | 1 µm filtered | unfiltered | 0.45 µm filtered |
| mg C/m² at 25 % flux decline | 90 | 127 | 50 | 70 |
| % of initial flux at 250 mg C/m² | 10 | 58 | 13 | 40 |

In summary, the stirred cell experiments with effluent isolates support the hypothesis that organic colloids play a major role in the fouling of low-pressure membranes. For both loose microfiltration membranes, the hydrophobic GVHP and the hydrophilic GSWP, the flux decline curves of the colloids (filtered as well as unfiltered) exhibit a more distinct fouling than either the hydrophobic acids (HPO-A) or the transphilic acids (TPI-A) fraction. This underlines the importance of colloids when investigating membrane fouling behaviour. In the "standard" XAD-8/ XAD-4 isolation procedure, as described by Aiken et al. [1992], the largest part of the colloidal material is lost in the non-adsorbable fraction (also-called hydrophilic or HPI fraction) which is seldom studied due to its high salt content in addition to its low organic carbon concentration. However, the isolation of the colloidal material using rotary-evaporation has its draw-backs too because an agglomeration of macromolecules during rotary-evaporation may occur and the re-dissolution of freeze-dried colloidal isolates is difficult. Nevertheless, this procedure allows for a better characterisation and understanding of the most severe foulants in low-pressure membrane filtration of wastewater treatment plant effluents.

# Chapter 6        Behaviour of foulants in membrane bio-reactors

## 6.1    Identification of main foulants

Two membrane bio-reactors (MBRs) are studied over a period of one year from December 2002 to November 2003. Samples are taken every two weeks from the membrane reactors. The activated sludge from the membrane reactor is separated from the water sample through a pre-filtration step (black ribbon paper filters from Schleicher and Schuell, Germany, with an approximate separation range of 12-25 μm; see Section 3.2 and 3.3) prior to the LC-OCD analysis. This enables the analysis of the dissolved/colloidal microbial products. Additionally, permeate samples are used for comparison purposes (pore size of hollow-fiber membranes in MBR pilot plants is 0.1-0.2 μm). Figure 6.1 depicts representative organic carbon chromatograms (left) and $UV_{254}$ absorption chromatograms (right) of a sample from the membrane reactor (filtrate) and the permeate for pilot plant 2. The main difference is the absence of the polysaccharides (PS, after 40 min) peak in the permeate. Thus, large macromolecules such as organic colloids, polysaccharides and proteins with molecular sizes of roughly 50,000 to > 150,000 D are retained by the membrane. On the other hand, smaller macromolecules, i.e. humic substances, their hydrolysates, and organic acids, are able to pass the membrane barrier. Only small amounts of these molecules are rejected by the membrane as can be seen in Figure 6.1 by the slightly smaller peak height in these peaks for the permeate sample. These molecules are likely trapped in the fouling layer or adsorbed onto activated sludge flocs during the membrane filtration. Hence, organic colloids and polysaccharides can be seen as main foulants in membrane bio-reactors. During the twelve month monitoring period of the two pilot plants, the evolution of the PS peak in the organic carbon chromatograms is investigated and correlated with operating conditions such as sludge retention time (SRT) or temperature.

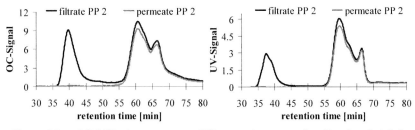

**Figure 6.1**    LC-OCD chromatograms of filtrate and permeate for pilot plant 2; left for organic carbon content, right for UV absorption at 254 nm

Besides the molecular size distribution of the organic carbon, the total organic carbon (TOC) of each sample is measured in the by-pass peak (see Section 3.5 for details). Figure

6.2 shows the evolution of both TOC and organic carbon content of the PS peak for the filtrate samples of pilot plant 1. A clear evolution over the year can be seen for both, with higher values during winter and lower values in summer. In general, the TOC and PS evolutions follow the same pattern. This suggests that an increase in total organic carbon is primarily due to an increase in large macromolecules (organic colloids, polysaccharides, and large proteins) as detected in the PS peak. In comparison, the humic substances and organic acids peaks in the LC-OCD chromatograms remain constant throughout the year. This is in accordance with the hypothesis that EfOM consists of natural organic matter (e.g. humic substances) from the drinking water and extracellular polymeric substances (EPS) produced by the mircoorganisms during the biological wastewater treatment process. The organic carbon content present in both peaks together, humic substances peak and organic acids peak, is also depicted in Figure 6.2 (HS + acids). The mean value for the humic substances plus organic acids is $11.0 \pm 1.2$ mg C/L. The dissolved organic carbon of Berlin tap water amounts to 4-5 mg/L exhibiting only the humic substances and organic acids peaks. Thus, humic-like substances in the wastewater stem, on one hand, from the drinking water source. On the other hand, they are also added to the wastewater by the consumer (e.g. faeces).

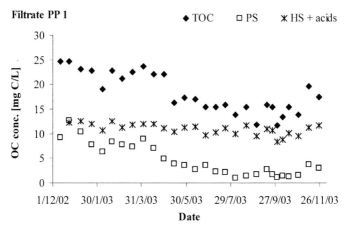

**Figure 6.2**    Evolution of total organic carbon (TOC), polysaccharides and organic colloids (PS), and humic substances and organic acids (HS + acids) over a 12 month period for the filtrate of pilot plant 1

Pilot plant 2 exhibits a similar trend showing a good correlation between TOC and PS (Appendix A, Figure A.1) while the mean value of humic substances plus acids amounts to exactly the same value as for pilot plant 1 ($11.0 \pm 1.2$ mg C/L). Thus, a difference in the performance of the two pilots plants can be fully attributed to the amount of organic colloids, polysaccharides, and large proteins as far as dissolved/colloidal molecules are concerned.

Since the large macromolecules such as polysaccharides, proteins, and organic colloids are retained completely by the MBR, the organic carbon in the permeate of both pilot plants mimics the course of the humic substances and organic acids in the filtrate samples. The total organic carbon concentration in the permeate is, therefore, stable at around $12.5 \pm 1.2$ mg C/L for pilot plant 1 and $12.6 \pm 1.6$ mg C/L for pilot plant 2, regardless of the incoming wastewater. The difference of these values to the humic substances and organic acids concentrations given above for the filtrate sample is due to amphiphilic compounds which elute after the organic acids peak.

Furthermore, the analysis of the by-pass peak in conjunction with the chromatograms reveals information on two more fractions: i) the total organic carbon rejected by the membranes and ii) hydrophobic compounds such as lipids which adsorb onto the SEC resin and therefore, are not accessible to an evaluation using size exclusion chromatography. For the first fraction (i), the difference between filtrate and permeate samples (=by-pass $TOC_{filtrate}$ – by-pass $TOC_{permeate}$) is calculated as it equals the total organic carbon retained by the membranes. Of the total organic carbon present in the filtrate samples an average of 4.6 mg C/L are retained in PP 1 and an average of 4.8 mg C/L in PP 2. On the other hand, the hydrophobic compounds (fraction ii) are assessed by comparing the TOC measured in the by-pass mode and the sum of the peaks in the chromatogram ($CDOC^6$). The hydrophobic compounds would be equal to the difference between by-pass TOC and CDOC. However, no significant amounts of hydrophobic compounds could be detected for the Ruhleben wastewater in both pilot plants (TOC and CDOC values are approximately identical within the measuring error of the instrument).

## 6.2    Influence of operational parameters

Pre-trials to this research in August 2002 are showing a higher polysaccharides peak for pilot plant 2 as depicted in Figure 6.3. This is in accordance with the higher fouling of the membrane in pilot plant 2 (post-denitrification) at that time. One of the objectives of the MBR pilot plant study is, therefore, to find out whether the pre-denitrification process design induces less fouling of the membrane than the post-denitrification or vice versa. However, by December 2002, when regular sampling started on both pilot plants for this research, the situation has been reversed with pilot plant 1 (pre-denitrification) exhibiting higher fouling and higher polysaccharides concentrations. Hence, a more important fouling potential can not be linked solely to either one of the two configurations tested, i.e. pre-denitrification (PP 1) or post-denitrification (PP 2), as apparently other parameters are of importance, too.

---

[6] CDOC = dissolved organic carbon accessible by size exclusion chromatography

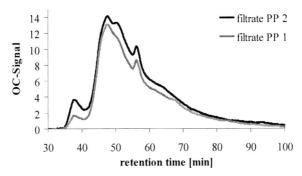

**Figure 6.3**     Organic carbon chromatograms (column HW-50S) of filtrate samples from
27th August 2002 for pilot plant 1 (bottom) and pilot plant 2 (top)
[Schumacher 2002]

During this research the pilot plants have been operated at two different sludge retention
times (SRT): an SRT of eight days from January until June 2003 and an SRT of fifteen
days from July until November 2003. In December 2002, the SRT has been decreased
continuously to reach the eight days by January 2003 as earlier research has been done
between 20-26 days sludge retention time [Gnirss et al. 2003a and 2003b]. A large amount
of fibers has been cut out of the membrane unit in pilot plant 2 for autopsy purposes on
15th September 2003. Aeration problems in the aerobic zone of pilot plant 2 have occurred
in October and November 2003 leading to a complete failure of nitrification from 23rd Oct.
- 7th Nov. 2003. Hence, only data from 8th January 2003 until 16th September 2003 are
taken into consideration for the following analysis and evaluation.

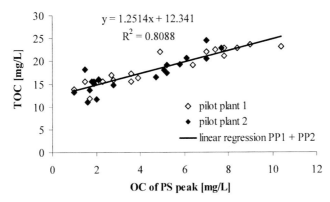

**Figure 6.4**     Correlation between TOC and organic carbon (OC) of PS peak in the filtrate
samples for both pilot plants

For this timeframe, the average overall correlation between the TOC of the permeate and the TOC of the filtrate has been: $TOC_{filtrate} = 1.25 * PS_{filtrate} + TOC_{permeate}$ (see Figure 6.4). Thus, the substances eluting in the PS peak account for 80 % of the TOC that is retained by the membrane. The remaining 20 % correspond to the rejection of small amounts of humic substances, organic acids, and amphiphilic compounds.

Since the compounds of the PS peak (organic colloids, polysaccharides, proteins) are the only ones being completely retained in the membrane bio-reactors, a good correlation with the fouling rate should be found. The fouling rate is defined here as the slope of the temperature corrected ($T_{ref} = 20°$ C) membrane resistance (for a more detailed description on how to calculate the fouling rate see Rosenberger et al. [2004]). Figures 6.5 and 6.6 depict the correlation between fouling rate and polysaccharides concentration measured photometrically as glucose equivalent by Dr. S. Rosenberger [Rosenberger et al. 2004]. The photometric measurements have been done in parallel to the LC-OCD analysis but samples are analysed weekly instead of every two weeks as with the LC-OCD. Hence, clearer correlations can be extracted from the photometric analyses due to more data points. A correlation between fouling rate and polysaccharides concentration in the filtrate samples of both pilot plants is only discernable for a sludge retention time (SRT) of eight days (Figure 6.5), while for a sludge retention time of 15 days only a poor correlation can be found, i.e. scattered points (see Figure 6.6).

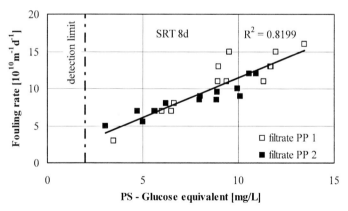

**Figure 6.5**    Correlation between fouling rate and polysaccharides (PS) measured photometrically as glucose equivalent [Rosenberger et al. 2004] for a sludge retention time of eight days

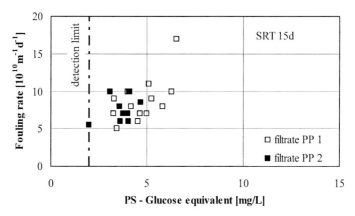

**Figure 6.6**     Correlation between fouling rate and polysaccharides (PS) measured
photometrically as glucose equivalent [Rosenberger et al. 2004] for a sludge
retention time of fifteen days

However, for an SRT of 15 days, all values (polysaccharides concentrations as well as
fouling rate) are in the low range and therefore, can have a higher measurement uncertainty
(detection limit 2 mg/l as glucose equivalent and less sensitivity of the photometric method
used by Rosenberger in the low concentration range). Additionally, the data points for an
SRT of 8 days span a period of six months, while only three months of sampling are
performed for an SRT of 15 days. This could be of importance as it can take several
months before stable operating conditions are achieved after a change of one or more
parameters such as the sludge retention time, i.e. the biomass needs time for adaptation.
Figure 6.7 depicts the correlation between fouling rate and the organic carbon content of
the PS peak as measured with the LC-OCD for an SRT of 8 days. Although fewer data are
available the trend is the same as with the photometric polysaccharides measurements.
Similar results are found for an SRT of 15 days, i.e. a poorer correlation between the
organic carbon (OC) content of the PS peak as measured with the LC-OCD and the fouling
rate is observed compared to an SRT of 8 days.

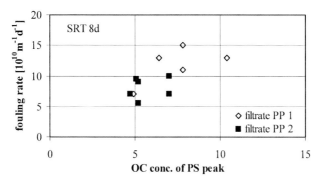

**Figure 6.7**    Correlation between fouling rate and organic carbon content of the PS peak
as measured with the LC-OCD for an SRT of 8 days

In their review of soluble microbial products (SMP), Barker and Stuckey [1999] state that
the amount of SMP produced is influenced by sludge retention time, organic loading rate,
and temperature. For sludge retention time and organic loading rate an optimal operating
point/ range appears to exist depending on the specific system (aerobic, anaerobic) and the
wastewater characteristics.

Nevertheless, a clear evolution of the PS peak in accordance with the change of SRT from
8 days to 15 days can not be found (see Figure 6.8): the PS peak starts to decrease from
nearly 8 mg C/L in April 2003 to around 2 mg C/L by June 2003. During this time the SRT
is held constant at approximately 8 days. It is only changed at the end of June 2003,
reaching a constant SRT of 15 days by mid July 2003.

Besides the sludge retention time, temperature plays an important role in the production of
extracellular polymeric substances (EPS)/ soluble microbial products (SMP). In general, an
increase in SMP is seen with decreasing temperature [Barker and Stuckey 1999]. For the
two MBR pilot plants in Ruhleben the temperature evolution of the wastewater
corresponds much better to the evolution of the PS peak than the sludge retention time.
Simultaneously to the decrease of the PS peak an increase in the wastewater temperature
from 17° C in mid April 2003 to 23° C in June 2003 is observed (see Figure 6.8 for pilot
plant 2; pilot plant 1 shows similar trends, see Appendix A, Figure A.2). Other operational
parameters such as MLSS and F/M ratio exhibit no clear correlation with the fouling rate
in the range investigated during this research [Rosenberger et al. 2004].

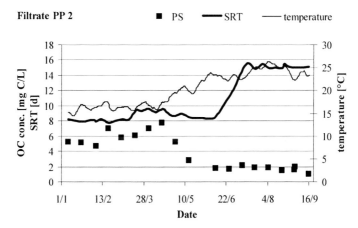

**Figure 6.8**    Evolution of PS peak (as organic carbon concentration), sludge retention time (SRT in days), and wastewater temperature (as 7-day running average) from 8.1.-16.9.2003 for pilot plant 2

Although no direct measurement has been done on the amount of stress the microorganisms have been exposed to, environmental stress turned out to be the most important factor in correlation to the PS concentration. Two occurrences lead to this impression:

1) The weekend of 7th/8th December 2002, one third of the sludge of pilot plant 1 has been lost resulting in an overload of organic carbon and nutrients for the remaining microorganisms in conjunction with a sudden decrease of the sludge retention time. Unfortunately, the first routine sampling of PP 1 has been on 11th December 2002. At this sampling, the PS concentration in the filtrate PP 1 sample is 4.5 times the PS concentration in filtrate PP 2. It takes nearly four months for the PS concentration to return to the same level as PP 2 (Figure 6.9).

2) In October 2003, pilot plant 2 had aeration problems in the aerobic zone which eventually lead to a complete loss of the nitrification process. A strong and immediate increase in the PS concentration is detected simultaneously with the decrease of dissolved oxygen in the aerobic reactors. Once the problem is fixed at the beginning of November, the PS concentrations decrease again.

Furthermore, it is not clear how the biological phosphorous removal within the anaerobic reactor impacts the production and degradation of the EPS. More in-depth studies would be of interest to resolve this issue.

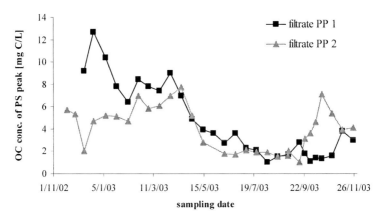

**Figure 6.9**     Evolution of the organic carbon concentration in the PS peak as measured
with the LC-OCD for all filtrate PP 1 and filtrate PP 2 samples

In summary, the following operating conditions have been found to be influential for the
fouling behaviour of the two membrane pilot plants:
-   stress situation for the microorganisms,
-   temperature, and
-   sludge retention time
regardless of the nitrification/denitrification process configuration.


## 6.3     Composition of the polysaccharides peak

### 6.3.1   Molecular size of PS peak

One characteristic of the polysaccharides and organic colloids as present in the PS peak is
the molecular size of these compounds. The molecular size can be assessed from the LC-
OCD analyses in conjunction with the stirred cell experiments/ pilot plant monitoring.
Depending on the calibration standards used, the HW-55S column of the LC-OCD can
separate molecules between a few hundred Dalton and approximately 200,000 D (see
Section 3.5). An exact correlation between molecular weight (D or g/mol) and molecular
size does not exist, but a rough approximation is that 100,000 D correspond to 10 nm.
Since the PS peak elutes very close to the void volume it cannot be ruled out that the
compounds of this peak have larger molecular sizes/ weights than the separation range.
This is supported by the shape of the peak which appears to have a right-skewed
distribution. Hence, the molecular size of the compounds eluting in the PS peak is at least
15-20 nm according to the separation range of the column (HW-55S). Secondly, it can be
seen from the stirred cell experiments and pilot plant monitoring that part of the
polysaccharides are retained by the membranes. If these compounds are retained by the
microfiltration membranes by sieving then a molecular size of more than 100-200 nm

could be expected (pore size of the membranes: 0.1 - 0.2 µm). To get a better idea of the molecular weight range of these substances, other SEC columns have to be used, i.e. spanning separation ranges up to a few million Dalton.

## 6.3.2   Comparison with extracellular polymeric substances (EPS)

In order to support the hypothesis that the polysaccharides are produced during the biological wastewater treatment process some additional analyses are performed. The aim is to see whether extracellular polymeric substances (EPS) that are extracted from the biomass of the activated sludge have the same organic carbon distribution as the paper filtered samples of the pilot plants (filtrate PP 1 and filtrate PP 2) when injected into the LC-OCD. In the literature, several methods are described to extract extracellular polymeric substances (EPS) from microorganisms [Rosenberger 2003, Späth 1998]. Oftentimes, chemical extraction is used in conjunction with centrifugation. The methods differ in the strength of extraction and possible damage caused to the bacterial cells which is undesirable because lysis products would then be measured as EPS [Rosenberger 2003].

An activated sludge sample from the membrane reactor of pilot plant 2 (sampling date: 7. Oct. 2003; MLSS concentration: 12.5 g/L) is used for the analyses. The EPS is extracted from the biomass of the activated sludge sample by ion-exchanger which is separated from the extracted EPS by sedimentation and centrifugation after 2 hours. The underlying principles is that by exchanging the $Ca^{2+}$ ions bridging the EPS to the cell wall, the EPS can then be removed physically from the cell walls by centrifugation (experiments are carried out by Dr. Sandra Rosenberger using the method described in Rosenberger [2003] without washing of the sludge). Parallel to the EPS extraction from the biomass, a filtrate PP 2 sample is mixed with the ion-exchanger (filtrate PP 2 + IX) as well as a sample of pure water (=reference sample).

The results of the LC-OCD analysis for the three samples are depicted in Figure 6.10 (organic carbon response) and Figure 6.11 (UV absorption at 254 nm). The chromatograms of the reference sample ($H_2O$ + IX) represent the amount of organic carbon and $UV_{254}$ absorption, respectively, caused by bleeding of the ion-exchanger as pure water itself gives no organic carbon or UV signal. The filtrate PP 2 + IX sample is identical to the filtrate PP 2 sample (filtrate PP 2 as measured in the routine analyses, i.e. without addition of ion-exchanger). The higher HS peak and acid peak and the additional peak at 82 min retention time for the filtrate PP 2 + IX sample compared to the filtrate PP 2 sample are due to the bleeding of the ion-exchanger (see reference sample, Figure 6.10).

The most important organic carbon peak for the extracted EPS sample (sludge + IX, Figure 6.10) is the PS peak. Hence, polysaccharides and organic colloids make up a large part of the extracellular polymeric substances. Proteins, on the other hand, do not seem to play such an important role as there is no clear peak around 53 min retention time where the bovine serum albumin (BSA) standard elutes. Of course, this can also mean that the proteins are larger than BSA and are eluting in the PS peak. Although the total organic carbon concentrations as measured in the by-pass peak differ significantly (Table 6.1), the organic carbon distribution is similar for the extracted EPS (sludge + IX) and filtrate PP 2 (with and without addition of IX). This is not the case for the $UV_{254}$ absorption chromatograms where the first peak of the extracted EPS clearly dominates the whole chromatogram (Figure 6.11). The first peak in the UV chromatogram is usually attributed to inorganic colloids [Huber and Frimmel 1996] although it is not clear where these would stem from in this case. Other possible explanations would be i) that the $UV_{254}$ absorption is caused by organic colloids present in the PS peak of the organic carbon chromatogram by light scattering or ii) that proteins are present in the PS peak of the organic carbon chromatogram which absorb UV light meaning that the two peaks do correspond or iii) a combination of the two aforementioned points. Further research is needed in this area to investigate the cause of the UV absorption and whether the first peak of the OC and UV chromatograms correspond to one another or not.

**Table 6.1**     TOC (by-pass peak) of EPS extraction samples

|              | activated sludge + IX | filtrate PP 2 + IX | filtrate PP 2 | $H_2O$ + IX |
|--------------|-----------------------|--------------------|---------------|-------------|
| TOC [mg/L]   | 130                   | 25.6               | 19.9          | 8.3         |

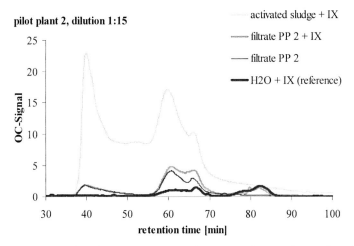

**Figure 6.10**     Organic carbon chromatograms of EPS extraction experiments (legend from top to bottom corresponds to chromatograms from top to bottom)

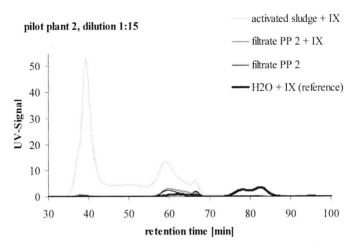

**Figure 6.11**    UV$_{254}$ absorption chromatograms of EPS extraction experiments (legend from top to bottom corresponds to chromatograms from top to bottom)

### 6.3.3  Sugar analyses

Furthermore, two samples are given to other institutes for sugar analysis. Investigations in the group of Prof. Kroh (Department for Food Chemistry, Technical University Berlin, Germany) are done with HPLC and thin-film chromatography. A filtrate PP 2 sample is analysed several times after concentration with rotary-evaporation by a factor of 500-1000. Due to the complex matrix of wastewater it is difficult to get clear results with the thin-film chromatography. However, the HPLC results show that most of the polysaccharides in the filtrate PP 2 sample are extremely large molecules as only few single sugars or simple polysaccharides (e.g. disaccharides) are detected. This supports the findings of the LC-OCD that the molecular weight of the polysaccharides is somewhere between 150,000 D and > 200,000 D (see Section 6.3.1). The highest peak detected by the HPLC of Prof. Kroh is anhydroglucogen (AHG) which is a sugar typically found in systems with microbiological activity.

The group of Prof. Croué (Laboratory for Water and Environmental Chemistry, University of Poitiers, France) analyses sugars and especially aminosugars using Py-GC-MS (pyrolysis gas chromatography with mass spectrometry). A sample of the isolated PS peak of a filtrate PP 2 sample is analysed. The PS peak is isolated using the LC-OCD by simply collecting the effluent of the UV detector between 33 min and 50 min retention time. Due to the dilution of the PS peak sample by the eluent of the LC-OCD, this sample has to be concentrated by a factor of 12 using rotary-evaporation. The PS peak of the filtrate PP 2 sample contains glucosamines and four amino acids (histidine, threonine, alanine, and valine), see Table 6.2. Glucosamines are of microbiological origin and can be found in chitin, glycolipids, mukopolysaccharides, and glycoproteins. N-acetyl-glucosamine is, for example, part of bacterial cell walls. The detected amino acids indicate the presence of proteins in the PS peak in addition to polysaccharides.

**Table 6.2**    Sugar and amino acids detected in the PS peak of filtrate PP 2 (values are corrected for the twelvefold concentration)

|  | nM |
| --- | --- |
| glucosamine | $2543 \pm 765$ |
| histidine | $8.4 \pm 1.2$ |
| threonine | $8.4 \pm 1.5$ |
| alanine | $2.8 \pm 0.3$ |
| valine | $1.1 \pm 0.4$ |

Hence, the analyses for sugars of the filtrate sample and its PS peak support the hypothesis that the compounds eluting in the PS peak are part of the EPS produced by the microorganisms during activated sludge treatment. This is in accordance to the findings of Debroux [1998]. He analysed effluent organic matter (EfOM) from different wastewater treatment plants and compared its characteristics to natural organic matter (NOM) isolates (isolates are obtained using the XAD-8/ XAD-4 isolation procedure of Aiken et al. [1992]). From the results of $^{13}$C-NMR and elemental analysis, he concludes that EfOM resembles microbially derived NOM as, for example, present in Lake Fryxell in Antartica where the humic substances are produced autochthonously (by algae) with no allochthonous NOM input. The standard reference isolate, Suwannee River Humic Substances, has much lower N/C ratios and a higher aromatic carbon content than EfOM. It can, thus, be assumed that organic biopolymers originating from the activated sludge biomass make up a large part of EfOM. However, Debroux [1998] reports also that an additional exposure of the EfOM to microbiota results in the humification of the EfOM because aliphatic structures are preferentially attacked by the microorganisms. This could be a possible reason for lower PS concentrations and lower fouling rates at higher sludge retention times.

## *6.4    Stirred cell experiments*

In addition to the routine samples taken every two weeks, stirred cell tests have been performed with filtrate and permeate samples of the two pilot plants. The aim is to confirm the fouling behaviour found in the pilot plants. For the stirred cell experiments 1-2 L of filtrate/ permeate are taken using the same procedure as for the routine samples (see Section 3.3). The membrane is the VVLP flat sheet membrane which comes closest to the Memcor hollow-fiber membrane installed in the pilot plants, i.e. a hydrophilized polyvinylidene fluoride microfiltration membrane with a pore size of 0.1-0.2 μm. All tests are conducted at a constant pressure of 0.3 bar (see Section 3.1). The initial flux $J_0$[7] varies

---

[7] $J_0$ is the flux measured after 1 min of sample filtration [L h$^{-1}$m$^{-2}$]

between the membrane specimens from 400 L/hm² to 750 L/hm². Hence, the normalized flux $J/J_0$ is used to compare experimental results.

Figure 6.12 shows the flux decline over the DOC delivered[8] to the membrane surface for the permeate and filtrate of pilot plant 2. It is clear that the filtrate sample has a high fouling potential. After a delivered DOC of 650 mg C/m², the flux has decreased to 20 % of the initial flux whereas the permeate sample has not fouled the membrane at all ($J/J_0$=1 at 650 mg C/m²). This strong flux decline of the filtrate occurs within the first 15 min of the stirred cell experiment. In the remaining time (total filtration time = 90 min) a further decline of 12 % occurs (final flux = 8 % of $J_0$) after a total of 1580 mg C/m² are delivered to the membrane surface. In comparison, for the permeate sample nearly 2000 mg C/m² are delivered to the membrane surface within the first 15 min of filtration causing only an overall flux decline of 7 %.

The difference between the filtrate PP 2 sample (taken 14. May 2003) and the permeate PP 2 sample (taken 16. April 2003) is that the permeate sample has passed through the hollow-fiber membrane of the MBR pilot plant. Thus, the permeate sample collected during the stirred cell experiment of permeate PP 2 has been filtered twice through the same type of membrane as the feed (permeate PP 2), having already undergone a first filtration in the pilot plant. Since the permeate PP 2 sample does not significantly foul a similar membrane, virtually all of the foulants must have been retained in the MBR. When comparing the LC-OCD chromatograms it becomes clear that the difference between the two feed waters is the PS peak which is only present in the feed of the sample "filtrate PP 2" and not in the feed of sample "permeate PP 2" (Figure 6.13). The corresponding permeate samples as collected during the stirred cell experiments give important information on the foulants. Permeate samples collected during the stirred cell experiments will be referred to as "perm" samples to avoid any confusion with the permeate of the MBR pilot plants (= permeate PP 1 or permeate PP 2). For the test performed with permeate PP 2, the chromatograms of feed and perm mimic each other with the only difference being that the perm chromatogram is slightly smaller in all areas than the feed chromatogram (Figure 6.13 right). Hence, no specific compound is preferentially retained by the membrane. Instead, the overall organic carbon load is responsible for the slight flux decline exhibited during the stirred cell test, i.e. humic substances, organic acids, and low molecular weight substances.

Although these substances are much smaller than the membrane pore size of $0.1 - 0.2$ μm, it is normal that small amounts are rejected by the membrane. This is due to the fact that the nominal pore size is never a 100 % cut-off as it is assumed for an ideal membrane. In fact, the correlation between the permeation coefficient and the molecular weight/ size is a continuous function [Eberle et al. 1979].

---

[8] delivered DOC [mg C/m²] = DOC$_{feed}$ [mg C/L] * permeate volume [L] / membrane surface area [m²]

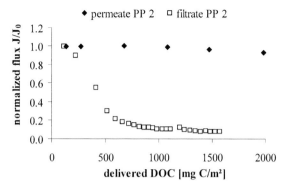

**Figure 6.12**    Flux decline of permeate and filtrate from pilot plant 2 in stirred cell tests
with hydrophilized PVDF microfiltration membranes (VVLP)

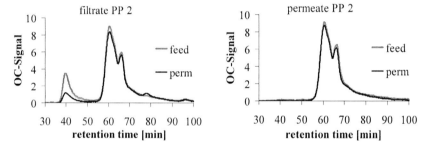

**Figure 6.13**    LC-OCD chromatograms for stirred cell test with filtrate PP 2 (left) and
permeate PP 2 (right); "feed" means the corresponding sample used for the
filtration test while "perm" is the permeate collected during the stirred cell
test

The feed and perm chromatograms for the stirred cell test performed with filtrate PP 2 are
depicted in Figure 6.13 (left). In this case, a clear difference between the feed and perm
chromatograms is found in the PS peak around 40 min retention time. The humic
substances and acid peaks are only diminished to a small extent as also seen in the pilot
plants' results. In contrast to the pilot plants' behaviour, the PS peak is not completely
retained by the membrane in the stirred cell test. One third of the initial PS concentration is
transported through the flat sheet membrane (VVLP) and is found in the perm sample.
Thus, from the large macromolecules (organic colloids, polysaccharides, and proteins),
0.8 mg C/L of the initial 2.4 mg C/L are most likely truly dissolved and smaller than the
membrane pore size of 0.1-0.2 μm.

Nevertheless, the PS peak is completely removed by the MBR pilot plants. Because the filtrate and permeate samples from the pilot plants are always taken in the middle of a filtration cycle (7 min after the last backwash), these samples might not be an appropriate comparison for the stirred cell experiments with clean unfouled membranes. In order to check this, permeate samples are collected from pilot plant 2 during a filtration cycle (10 s, 30 s, 60 s, 2 min, 3 min, 10 min, and 12 min after the last backwash on 4. Nov. 2003) and each sample is analysed with the LC-OCD. However, none of the samples exhibits a PS peak and no evolution of the TOC (as measured in the by-pass peak) over the filtration time is found. A second sampling campaign of a subsequent filtration cycle (sampling date: 4. Nov. 2003) gives the same results suggesting that the permeate quality does not depend on the filtration time. Unfortunately, a similar sampling campaign could not be done directly after a chemical cleaning of the membrane in the pilot plants which would have come closer to the state of a clean unfouled membrane compared to that after a backwash.

This means that the stirred cell experiments do not accurately replicate the fouling situation in membrane bio-reactors for most of the time (a possible exception would be directly after a chemical cleaning). Nevertheless, stirred cell experiments do give a qualitative indication of the fouling potential and behaviour. The difference between the stirred cell tests and the pilot plants is the presence of suspended biology, i.e. the activated sludge, in the latter. This means that the activated sludge plays an important role in the rejection characteristics of the membranes since a paper filtered sample from the membrane reactor of the pilot plants (=filtrate) is being used as feed water for the stirred cell tests. Possible explanations for the complete rejection of the PS peak, including any truly dissolved compounds with a molecular size smaller than the pore size of the membrane, are:
  - the formation of a layer of activated sludge flocs on top of the membrane surface acting as an additional filter entrapping any dissolved large macromolecules,
  - the formation of a fouling layer including activated sludge flocs and large colloidal macromolecules forming a secondary membrane and thereby, reducing the nominal pore size of the microfiltration hollow-fiber membrane,
  - the formation of a biofilm on the membrane surface that feeds on the dissolved macromolecules (according to Huber and Gluschke [1998] substances eluting in the PS peak are biodegradable although these compounds are also produced by bacteria in the form of extracellular polymeric substances).

With regards to the fouling rate of the pilot plant membranes, the amount of organic carbon present in the PS peak plays a crucial role (see Section 6.2). Stirred cell experiments with filtrate PP 1, filtrate PP 2, and filtrate CAS (conventional activated sludge from WWTP Ruhleben, see Section 3.3) indicate that the specific characteristics of the compounds eluting in the PS peak play an important role in the fouling behaviour, too. Figure 6.14 depicts the flux decline curves of the three aforementioned samples with the VVLP membrane (hydrophilic MF, 0.1-0.2 μm, PVDF). While filtrate PP 1 and filtrate PP 2 show similar fouling behaviour, the filtrate CAS sample exhibits a lower flux decline. However, all three samples have similar TOC concentrations and nearly identical organic carbon concentrations of the PS peak (Table 6.3). The sludge retention time of all three systems is around 15 days at the time of sampling. Besides the operational difference, i.e. conventional activated sludge versus membrane bio-reactor, the water temperature in the conventional WWTP is lower at the time of sampling (filtrate CAS sampled 16.4.03) than in the membrane bio-reactor pilot plants (sampled 19.8.03). A lower temperature should induce higher fouling as more EPS is produced by the microorganisms [Barker and Stuckey 1999]. Besides differences in operational conditions as the cause for the different characteristics of the compounds eluting in the PS peak, the activated sludge microbiology differs somewhat between conventional systems and MBRs [Rosenberger 2003].

**Table 6.3**     Organic carbon concentrations of the bulk sample and PS peak for filtrate PP 1, filtrate PP 2, and filtrate CAS

|  | TOC [mg C/L] | OC of PS peak [mg C/L] |
|---|---|---|
| Filtrate PP 1, 19.8.03 | 14.6 | 1.4 |
| Filtrate PP 2, 19.8.03 | 16.3 | 1.3 |
| Filtrate CAS, 16.4.03 | 15.6 | 1.4 |

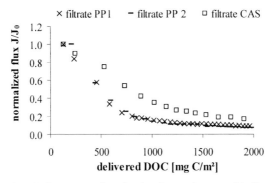

**Figure 6.14**     Flux decline curves for stirred cell experiments with filtrate PP 1, filtrate PP 2, and filtrate CAS with the VVLP membrane (hydrophilized MF)

Additional stirred cell experiments are carried out with filtrate PP 1 using either a microfiltration membrane (MF, VVLP) or an ultrafiltration membrane (UF, YM100). The results of these tests are plotted in Figure 6.15 as normalized flux versus delivered DOC. As can be seen, both membranes are fouled to the same extent after 90 min filtration (the initial flux has declined to less than 10 %). However, there seems to be a difference in the fouling behaviour with the microfiltration membrane fouling more slowly than the ultrafiltration membrane: 187 mg C/m² delivered DOC cause 40 % flux decline with the ultrafiltration membrane while 521 mg C/m² are necessary for the same flux decline with the microfiltration membrane.

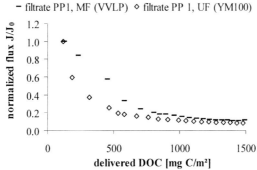

**Figure 6.15**   Flux decline curves of stirred cell tests for filtrate PP 1 with a microfiltration membrane (VVLP) and an ultrafiltration membrane (YM100)

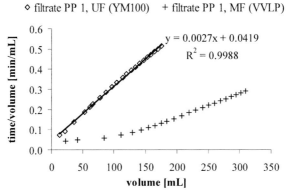

**Figure 6.16**   Identification of fouling mechanism for filtrate PP 1 with a microfiltration membrane (VVLP) and an ultrafiltration membrane (YM100); a linear correlation between time per volume and volume indicates cake formation as is the case for the UF membrane here (equation and $R^2$ value given in graph)

A linearization of the curves can help in identifying the underlying fouling mechanism. Linear plots of flux (J) versus filtration volume (V), t/V versus t (time), and t/V versus V indicate pore blockage, pore constriction, and cake formation, respectively, as dominant fouling mechanism (see Section 2.1.5). Figure 6.16 shows the graph for t/V versus V. The data of the ultrafiltration stirred cell experiment can be linearized using the t/V versus V plot and thus, the dominant fouling mechanism here is cake formation. The results with the microfiltration membrane, on the other hand, are not linearized in Figure 6.16 nor using one of the other two plots (data not shown). This means that either two or more fouling mechanisms are relevant, e.g. pore constriction and cake formation.

# Chapter 7    Statistical analysis and modelling of membrane fouling by municipal wastewater

Theoretical analyses of the fouling data obtained from stirred cell and pilot plant experiments are presented and discussed in this chapter. In a first step, a statistical method, principal component analysis (PCA), is employed to determine the parameters that are most influential on the fouling of low-pressure membranes (Section 7.1). The principal component analysis is applied to both the stirred cell experiments as well as the monitoring data from the two MBR pilot plants. In order to further investigate the mechanisms causing fouling in low-pressure membrane filtration of municipal wastewater, a mathematical model is applied to the results obtained with the stirred cell tests in a second step (Section 7.2).

## 7.1  Principal component analysis

The basic idea of principal component analysis (PCA) is the reduction of a multi-dimensional data set of interrelated variables while losing as little of the variation in the original data set as possible. In order to achieve this, the original data set is transformed into a new set of uncorrelated variables or principal components. Thus, the output of a PCA is a number of principal components which are ordered according to the amount of variation they account for. This implies that the first few principal components (usually the first two or three) account for most of the variation in the original data set [Jolliffe 1986].

The principal component analyses are performed using a software package, STATISTICA (StatSoft Inc., Tulsa, USA). The first step is to compile a meaningful database (the databases for the stirred cell experiments and the MBR pilot plants are presented in Appendix B). It must be mentioned that PCA can only be employed if the database is complete, i.e. without any "holes" due to missing data.

A correlation matrix, the so-called Pearson correlation matrix, is generated for the selected data set by simple-linear regression (least square regression). The Pearson correlation matrix contains a correlation coefficient (Pearson r value[9]) for each pair of variables in the data set. The correlation coefficient describes to which extent the two variables are linearly related and can assume values between $-1$ and $+1$. A value of $-1$ is found in the case of a perfect negative correlation while a value of $+1$ represents a perfect positive correlation. Hence, the closer r is to the number 1 (positive or negative) the clearer is the linear

---

[9] $r = \dfrac{n(\Sigma XY) - (\Sigma X)(\Sigma Y)}{\sqrt{\left[ n\Sigma X^2 - (\Sigma X)^2 \right]\left[ n\Sigma Y^2 - (\Sigma Y)^2 \right]}}$

relationship between the two corresponding variables whereas no linear relationship can be inferred if r is close to zero [StatSoft 2000].

Using the Pearson correlation matrix an eigenvector decomposition is performed to transform the data set of variables into a set of principal components [Jarusutthirak 2002]. The number of principal components is equal to the number of variables in the original data set. However, only the first two or three principal components are usually retained for interpretation since the first principal component accounts for most of the variation in the original data set, the second principal component accounts for most of the remaining variation and so forth.

An eigenvalue is attributed to each principal component. The eigenvalues describe the amount of variance the principal component accounts for in the original data set. The eigenvalues are used in the selection of the number of principal components to retain for interpretation. Several rules exist on how to determine the number of principal components that should be selected. The first possibility is to determine a cumulative percentage of total variation, e.g 80 %, which the retained principal components must contribute. Another possibility is to adhere to Kaiser's rule and to only retain principal components with an eigenvalue $\geq 1$. Furthermore, a so-called scree plot can be drawn with the principal component number as x-axis and the corresponding eigenvalues as y-axis. One has then to determine visually where the slope changes from 'steep' to 'not steep'. Any components to the left of that point, i.e. on the steep part of the slope, are retained [Jolliffe 1986]. In this study Kaiser's rule is used.

In order to interpret the results, component (or factor) loadings are assigned to the original variables. They relate each variable with each principal component. As for the correlation matrix a number close to $-1$ or $+1$ means that this variable influences the principal component strongly while variables with component loadings around zero have no significant influence.

Finally, principal component (or factor) scores can be calculated for each case in the original data set. A graphical presentation of these principal component scores using a two-dimensional abstract space (x-axis = principal component 1; y-axis = principal component 2) helps in identifying patterns or clusters within the original data set and thus, in finding the significant influences between variables and cases [Jarusutthirak 2002].

### 7.1.1  Results for stirred cell experiments

The aim of the principal component analysis with the experimental data gathered during stirred cell tests is to assess how membrane properties and feed water characteristics affect the flux decline trends observed during membrane filtration. Table 7.1 gives an overview of the different cases and variables used for the statistical analysis. Stirred cell experimental results with all five membranes (YM100, MX500, VVLP, GSWP, and

GVHP) are taken into consideration. The feed waters can be subdivided in bulk EfOM samples (from the WWTP Boulder or the WWTP Ruhleben), EfOM isolates (colloids, HPO-A, TPI-A), MBR samples (filtrate and permeate of the two pilot plants), and SMP samples. The SMP samples have been produced using a sequencing batch reactor (SBR) fed with glucose (see Jarusutthirak [2002] for details on the system). The variables include feed water characteristics (DOC, $UVA_{254}$, and SUVA), membrane properties (pore size, PWP, pressure), and fouling indices (delivered DOC at 25 % flux decline, normalized flux $J/J_0$ at a delivered DOC of 250 mg C/m² and 500 mg C/m², respectively).

**Table 7.1**    Details of cases and variables that are contained in the database for the stirred cell experiments (complete database in Appendix B, Tables B.1 and B.2)

| Cases | | Variables | | |
|---|---|---|---|---|
| feed water | membranes | water characteristics | membrane properties | fouling indices |
| - EfOM | - UF: YM100 | - DOC | - pore size | - delivered DOC at 25 % flux decline |
| - Colloids | - UF/MF: MX500 | - $UVA_{254}$ | - PWP | |
| - HPO-A | - MF: VVLP, | - SUVA | - pressure | - flux $J/J_0$ at 250 mg C/m² del. DOC |
| - TPI-A |    GSWP, | | | |
| - SMP |    GVHP | | | - flux $J/J_0$ at 500 mg C/m² del. DOC |
| - MBR | | | | |

The correlation matrix needed to perform a principal component analysis is given in Table 7.2. The Pearson r-values always correlate two of the variables with one another. An r-value of > 0.70 is indicative of a significant correlation between two parameters. Thus, membranes with larger pore sizes exhibit a higher pure water permeability although less pressure has been applied (positive correlation for pore size–PWP and negative correlation for pore size–pressure). The good correlation between the samples' DOC and $UVA_{254}$ values suggest a predominance of humic substances in most of the samples. This is supported by the size exclusion chromatograms (see Chapter 4-6). Furthermore, correlations exist between the delivered DOC at 25 % flux decline (del. DOC) and the remaining normalized flux after 500 mg C/m² are delivered to the membrane surface (flux @ 500), and between the two parameters 'flux @ 250' and 'flux @ 500'. However, no correlation is found between 'del. DOC' and 'flux @ 250'. This suggests that of the three fouling indices, the 'flux @ 500' is redundant in terms of the information contained in the other two parameters and could be omitted.

**Table 7.2**    Pearson correlation matrix

|  | Pore size | PWP | Pressure | DOC | UVA$_{254}$ | SUVA | Del. DOC | Flux @ 250 | Flux @ 500 |
|---|---|---|---|---|---|---|---|---|---|
| Pore size | 1.00 | **0.74** | **-0.70** | -0.09 | -0.08 | -0.23 | 0.08 | -0.14 | -0.07 |
| PWP |  | 1.00 | -0.45 | -0.34 | -0.23 | -0.14 | 0.19 | 0.03 | 0.08 |
| Pressure |  |  | 1.00 | -0.30 | -0.26 | 0.11 | -0.10 | 0.00 | 0.09 |
| DOC |  |  |  | 1.00 | **0.92** | 0.22 | -0.13 | 0.17 | -0.01 |
| UVA$_{254}$ |  |  |  |  | 1.00 | 0.49 | -0.11 | 0.22 | 0.04 |
| SUVA |  |  |  |  |  | 1.00 | -0.02 | 0.20 | 0.13 |
| Del. DOC |  |  |  |  |  |  | 1.00 | 0.59 | **0.74** |
| Flux @ 250 |  |  |  |  |  |  |  | 1.00 | **0.92** |
| Flux @ 500 |  |  |  |  |  |  |  |  | 1.00 |

On the basis of the Pearson correlation matrix, nine components are extracted. The number of initially extracted components is equal to the number of variables in order to ensure that no variation in the original data set is lost. An eigenvalue is assigned to each component (Table 7.3). The first three components have very similar eigenvalues between 2.1 and 2.6, each accounting for 24-29 % of the total variation in the original data set. The eigenvalues of component 4 and the following five components are below 1.0 and each explain less than 10 % of the variation in the original data set. Thus, in accordance to Kaiser's rule the first three components are retained for interpretation.

**Table 7.3**    Eigenvalues of extracted components

|  | Eigenvalue | % total variance | Cumulative % |
|---|---|---|---|
| Component 1 | 2.62 | 29.0 | 29.0 |
| Component 2 | 2.48 | 27.6 | 56.6 |
| Component 3 | 2.13 | 23.7 | 80.3 |
| Component 4 | 0.86 | 9.5 | 89.8 |
| Component 5 | 0.40 | 4.5 | 94.3 |
| Component 6 | 0.27 | 2.9 | 97.2 |
| Component 7 | 0.18 | 2.1 | 99.3 |
| Component 8 | 0.04 | 0.4 | 99.7 |
| Component 9 | 0.03 | 0.3 | 100 |

The component loadings of each of the three principal components are given in Table 7.4. Component 1 is influenced by the fouling indices: flux $J/J_0$ after 250 mg C/m² are delivered to the membrane surface (flux @ 250), and the flux $J/J_0$ after 500 mg C/m² are delivered to the membrane surface (flux @ 500). The first parameter exhibits a higher r-value suggesting that this fouling index has a higher sensitivity when comparing different stirred cell experiments. Component 2 represents the variations which can be accounted by pure water permeability (PWP) of the membrane specimens and the variations in delivered DOC when 25 % flux decline are observed (del. DOC). For both parameters a negative correlation means that larger values correspond to less fouling. For example, a higher pure water permeability is characteristic for membranes with larger pore sizes (see correlation matrix in Table 7.2) which reject generally less components. The delivered DOC at 25 % flux decline ($J/J_0 = 0.75$) was intended as a fouling index. However, it includes the feed water characteristics in terms of the indicated organic carbon content. This might be an explanation for the fact that none of the feed water characteristics (DOC, $UVA_{254}$, and SUVA) are represented in the first three principal components. Additionally, the organic carbon concentrations of the stirred cell experiments with HPO-A and TPI-A isolates have been set at 4-5 mg/L (10 mg of lyophilized isolate are resolubilized per litre ultra-pure water for the tests). Thus, there is little variation in the amount of DOC, although the DOC character as described by the SUVA varies quite significantly over the complete data set (between 0.57 L $mg^{-1}m^{-1}$ and 6.48 L $mg^{-1}m^{-1}$, see Appendix B, Tables B.1 and B.2). Finally, component 3 is defined by the pressure applied during the stirred cell tests. The pressure has been set arbitrarly at either 0.3, 0.6 or 1.0 bar depending on the membrane (MF/UF, hydrophilic/hydrophobic) used in a stirred cell experiment; hence, the correlation between pore size and applied pressure (see Table 7.2).

**Table 7.4**      Component loadings

|  | Component 1 | Component 2 | Component 3 |
|---|---|---|---|
| Pore size | -0.43 | -0.58 | -0.60 |
| PWP | -0.36 | **-0.72** | -0.32 |
| Pressure | 0.14 | 0.37 | **0.82** |
| DOC | 0.51 | 0.43 | -0.66 |
| $UVA_{254}$ | 0.58 | 0.40 | -0.67 |
| SUVA | 0.50 | 0.27 | -0.14 |
| Del. DOC | 0.48 | **-0.70** | 0.17 |
| Flux @ 250 | **0.81** | -0.47 | 0.08 |
| Flux @ 500 | **0.74** | -0.60 | 0.23 |

Figure 7.1 depicts the plot of the component scores for each stirred cell experiments. The two-dimensional plot is chosen (x-axis: component 1, y-axis: component 2) because component 3 is solely correlated to the applied pressure, i.e. the third dimension would merely arrange the stirred cell experiments according to the different types of membranes (1 bar: YM100 & MX500; 0.6 bar: GVHP; 0.3 bar: GSWP & VVLP).

Colloids and SMP samples induce the highest flux decline (i.e. smallest values for $J/J_0$) and are thus found on the far left in the graph. The fact that they both cluster in the same area of the diagram supports the hypothesis that colloids are to a large extent microbiologically derived. The other two isolated effluent organic matter fractions, HPO-A and TPI-A, induce similar flux decline on the different membranes as they cluster around the y-axis (zero for component 1). The observed variation in component 2 must be attributed to the differences in pure water permeability of the membranes as the feed water DOC has been adjusted to approximately 5 mg/L for each experiment with these isolates. The most variation is found for the Boulder EfOM and Ruhleben EfOM samples. On one hand, they scatter due to variations in the samples' DOC and in accordance to the different membranes assessed (component 2: delivered DOC and PWP). On the other hand, the variation on the x-axis (component 1), i.e. the amount of observed flux decline, suggests that the bulk EfOM fouls the membranes to different degrees depending on the pore size and the membrane material. Finally, the filtrate samples of the two MBR pilot plants cluster in the upper right quadrant while the MBR permeate sample induces the lowest flux decline (highest value for $J/J_0$) and thus, is found the farthest to the right in the graph. The DOC of the MBR permeate sample is 13 mg/L, i.e. approximately 2-3 times higher than the concentrations in the HPO-A and TPI-A isolate experiments (4-5 mg/L). Nevertheless, the isolates foul the membranes to an higher extent, based on the indicated fouling indices. This emphasizes the importance of the nature and characteristics of the organic matter in the assessment of membrane fouling. The MBR permeate sample, having already been filtered through an MF membrane (hydrophilized PVDF, pore size 0.1-0.2 µm), is free of any substances eluting in the PS peak (organic colloids, polysaccharides, large proteins) whereas the isolates exhibited a small PS peak in size exclusion chromatography (see Chapter 5, Figure 5.1). Hence, for the assessment of the fouling potential of a specific feed water it is of more importance to know the DOC distribution and especially the amount of organic matter eluting in the PS peak than the bulk organic carbon concentration.

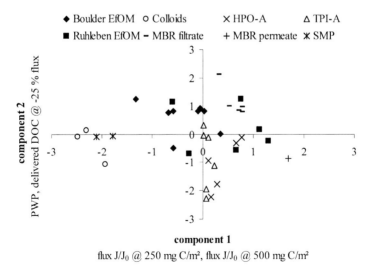

**Figure 7.1**    Plot of component scores for stirred cell experiments

## 7.1.2 Results for MBR pilot plants

In the principal component analysis of the MBR pilot plant data, water characteristics and operational parameters are used as variables (Table 7.5 and Tables B.3 and B.4 in Appendix B). The feed water is characterised with size exclusion chromatography and a turbidimeter (see Chapter 3 for details on the instruments). Four parameters are extracted from the resulting organic carbon and UV absorption chromatograms: the total organic carbon as measured in the by-pass peak, the organic carbon content of the PS peak, the sum of the organic carbon concentration present in the humic substances (HS) and organic acids peaks together, and the $UVA_{254}$ of the first peak eluting in the UV chromatogram (inorganic colloids). The sludge retention time and the feed water temperature are entered as variables for the operational parameters.

**Table 7.5**    Details of cases and variables that are contained in the database for the MBR pilot plant study (complete database in Appendix B, Tables B.3 and B.4)

|   | Cases | Variables | |
| --- | --- | --- | --- |
| pilot plant | sampling date: | water characteristics | operational parameters |
| - PP 1 | one sample every | - TOC | - SRT |
| - PP 2 | two weeks from | - OC (PS peak) | - temperature |
|  | Jan. 2003 until | - OC (HS + organic acids peak) |  |
|  | Nov. 2003 | - $UVA_{254}$ (inorganics peak) |  |
|  |  | - turbidity |  |

The aim of the statistical analysis is to investigate the influence of sludge retention time (SRT) and temperature on the feed water characteristics in pilot plant 1 and pilot plant 2. Which of the two parameters has the more significant impact on the amount of different organic compounds in the filtrate samples (paper filtered activated sludge from the membrane compartment of each pilot plant), especially on those substances eluting in the PS peak?

Thus, the first step is to calculate a correlation matrix. Table 7.6 summarizes the Pearson r-values correlating always two of the variables with each other. An r-value of $> 0.70$ is needed in order to assume a significant correlation between two parameters. This is the case for the parameter pairs TOC–PS, TOC–HS+acids, and PS–temperature. Hence, a good (positive) correlation exists between the total organic carbon and the organic carbon content of the PS peak as well as with the organic carbon content of the combined humic substances and organic acids peaks. This is not surprising since the TOC concentration has been shown to follow the evolution of the PS peak content over the year (see Figure 6.2). Although the organic carbon concentration of the HS plus organic acids peaks remains more or less stable throughout the year, it accounts for over 50 % of the total organic carbon in the filtrate samples which explains the good correlation between TOC and HS+acids.

**Table 7.6**     Pearson correlation matrix

|            | TOC  | PS   | HS+acids | $UVA_{254}$ | Turbidity | SRT   | Temperature |
|------------|------|------|----------|-------------|-----------|-------|-------------|
| TOC        | 1.00 | **0.88** | **0.84** | 0.66    | 0.39      | -0.53 | -0.57       |
| PS         |      | 1.00 | 0.68     | 0.60        | 0.35      | -0.67 | **-0.73**   |
| HS+acids   |      |      | 1.00     | 0.54        | 0.47      | -0.42 | -0.38       |
| $UVA_{254}$ |     |      |          | 1.00        | 0.54      | -0.22 | -0.49       |
| Turbidity  |      |      |          |             | 1.00      | 0.01  | -0.38       |
| SRT        |      |      |          |             |           | 1.00  | 0.46        |
| Temperature |     |      |          |             |           |       | 1.00        |

More interesting than these two correlations with the TOC, is the negative correlation between the PS peak and the feed water temperature. As seen in Chapter 6, the compounds eluting in the PS peak are major foulants of the hollow-fiber MF membranes installed in the MBR pilot plants. These compounds are at least partially produced during the biological wastewater treatment, i.e. by the microorganisms, in the form of extracellular polymeric substances (EPS). The correlation between the feed water temperature and the PS peak supports the hypothesis that the microorganisms are immediately responding to changes in their environment. Barker and Stuckey [1999] state that with lower temperature more soluble microbial products (SMP) are produced. Thus, low temperature (e.g. 15° C in the case of the two MBR pilot plants) can be seen as a stress factor for the microorganisms. In contrast, the sludge retention time (8 days versus 15 days) is not that influential on the amount of organic colloids and polysaccharides detected. This is in agreement with the findings of Jarusutthirak [2002] who found no clear differences in the characteristics of soluble microbial products for sludge retention times of 5, 10, and 30 days.

Furthermore, the non-existence of a correlation between the PS peak and $UVA_{254}$ of the first peak in the UV chromatogram is worth mentioning. Huber and Frimmel [1996] associate the first UV peak with inorganic colloids such as Fe, Al or Si colloids. This is in accordance with the absence of UV absorption by polysaccharides (absence of structures with de-localized $\pi$-electron systems necessary for UV absorption at 254 nm). However, organic colloids and proteins could have an impact on this peak. Proteins would exhibit UV absorption while organic colloids might scatter the UV light due to their size similar to the inorganic colloids. Thus, the missing correlation between the first peak (PS peak) in the organic carbon (OC) chromatogram and the first peak in the UV chromatogram means that either only inorganic colloids are responsible for the UV peak as suggested by Huber and Frimmel [1996] or the amount of organic colloids and proteins in the PS peak is too small in comparison to the non-absorbing polysaccharides to allow for a good correlation.

In a second step, principal components are extracted. The number of components extracted is identical to the number of variables used (i.e. seven for the MBR data) in order to ensure that no information is lost in the process of the statistical analysis. However, according to Kaiser's rule, only components with eigenvalues > 1.00 are meaningful in order to explain the variation in the original data set. Table 7.7 gives the eigenvalues for each of the extracted seven components and the percentage of total variance in the original data set that each component accounts for. Thus, the first two components that are retained for interpretation account for 76.5 % of the variation in the original data set.

**Table 7.7**     Eigenvalues of extracted components

|             | Eigenvalue | % total variance | Cumulative % |
|-------------|------------|------------------|--------------|
| Component 1 | 4.19       | 59.9             | 59.9         |
| Component 2 | 1.16       | 16.6             | 76.5         |
| Component 3 | 0.67       | 9.6              | 86.1         |
| Component 4 | 0.44       | 6.2              | 92.3         |
| Component 5 | 0.33       | 4.7              | 97.0         |
| Component 6 | 0.15       | 2.1              | 99.1         |
| Component 7 | 0.06       | 0.9              | 100          |

Table 7.8 presents the component loadings for component 1 and component 2. Unfortunately, component 1 is influenced by nearly all the variables, i.e the parameters representing organic carbon concentrations (TOC, OC of PS peak, OC of HS+acids peaks), $UVA_{254}$, and temperature. Only sludge retention time and turbidity are not significantly contributing to component 1. In contrast, component 2 is primarily influenced through the samples' turbidity.

**Table 7.8**     Component loadings

|             | Component 1 | Component 2 |
|-------------|-------------|-------------|
| TOC         | **-0.93**   | 0.05        |
| PS          | **-0.93**   | 0.23        |
| HS+acids    | **-0.82**   | -0.09       |
| $UVA_{254}$ | **-0.75**   | -0.39       |
| Turbidity   | -0.55       | **-0.70**   |
| SRT         | 0.62        | -0.67       |
| Temperature | **0.74**    | -0.10       |

The turbidity is measured from the paper filtered activated sludge samples taken in the membrane compartment of the two MBR pilot plants (filtrate PP 1 and filtrate PP 2 samples). The turbidity measurements are performed to assess the paper filtration. The paper filters (black ribbon, Schleicher & Schuell, Germany) have an approximate filtration range of 12-25 μm. However, the effective pore size is probably much smaller due to the immediate formation of a filtration cake by the activated sludge flocs. Differences in the turbidity could be due to variations in the paper filters or the filtration cake. In the case of the latter, this would indicate that the composition of the activated sludge has changed.

Figure 7.2 depicts the diagram of the principal component scores for all samples from pilot plant 1 and pilot plant 2 analysed between January and November 2003. Component 2 (turbidity) is plotted against component 1 (TOC, PS peak, HS+acids peak, $UV_{254}$, and temperature). The samples of both pilot plants are distributed similarly. This supports the finding that the configuration of the pilot plants, i.e. pre-denitrification versus post-denitrification, does not have a significant impact on the fouling behaviour. Instead, the feed water temperature must be seen as stress factor for the microorganisms inducing higher polysaccharides and organic colloids concentrations with lower temperature.

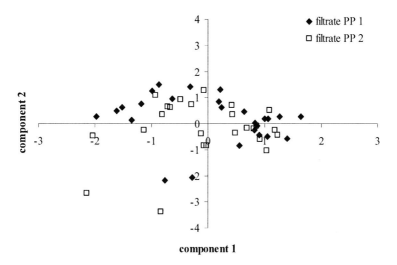

**Figure 7.2**      Plot of component scores for all samples from Jan.-Nov. 2003 (PP1 & PP2)

## 7.2    *Modelling of low-pressure membrane fouling*

As described in Section 2.1.2, replotting the flux decline data using characteristic coordinates can help in determining the predominant fouling mechanism. Figure 7.3 shows the graphs using the characteristic coordinates on a stirred cell test with Boulder EfOM and the loose hydrophilic UF membrane (YM100):

   a)  cake formation (filtration time/permeate volume versus permeate volume),
   b)  pore constriction (filtration time/permeate volume versus filtration time),
   c)  pore blockage (flux versus permeate volume).

If the shape of the resulting curves is linear in one of the graphs, one can attribute the observed flux decline to the corresponding fouling mechanism. For the depicted stirred cell test (Figure 7.3), this is the case in the t/V versus V plot and therefore, it is assumed that cake formation is the predominant fouling mechanism here. As can be seen from the graphs, a very high $R^2$ value of > 0.99 is needed in order to indicate the predominance of one of the classical fouling mechanisms.

a)

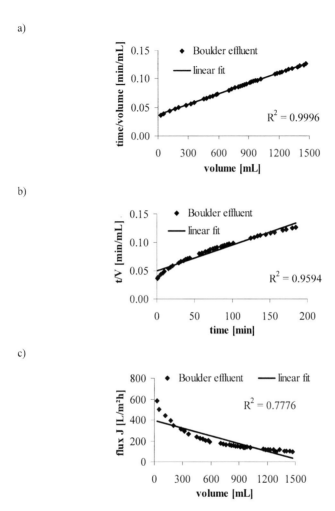

b)

c)

**Figure 7.3**    Flux decline data for Boulder EfOM with the loose hydrophilic UF
membrane (YM100) plotted in accordance with a) the cake formation
model, b) the pore constriction model, and c) the pore blockage model

Table 7.9 summarizes the results of all stirred cell experiments with regard to the $R^2$ values
obtained from linear fits in the characteristic plots for pore blockage, pore constriction, and
cake formation. The last column indicates the predominant fouling mechanism as given by
an $R^2$ value of $> 0.99$. In the case where none of the classical fouling mechanism plots
result in a linear relationship "---" is entered for the corresponding stirred cell experiment.
In most cases, where one of the classical fouling models can explain the flux decline data,

cake formation is the predominant fouling mechanism. A linear fit of the experimental data according to the pore constriction model is obtained for six stirred cell tests. However, at the same time, a linear fit according to the cake formation model is seen with a slightly higher $R^2$ value. Thus, it is not clear whether cake formation or pore constriction or a combination of both is the underlying fouling mechanism. In none of the stirred cell experiments can the fouling behaviour of the different feed waters be described by pore blockage.

**Table 7.9**   $R^2$ values for linear fits of the flux decline data using the characteristic coordinates for the pore blockage model, pore constriction model, and cake formation model

| Mem-brane | Feed water | Pore blockage ($R^2$-value) | Pore constriction ($R^2$-value) | Cake formation ($R^2$-value) | Predominant fouling mechanism |
|---|---|---|---|---|---|
| YM100 | Boulder EfOM | 0.8543 | 0.9776 | 0.9977 | cake formation |
| YM100 | Boulder EfOM | 0.7783 | 0.9574 | 0.9993 | cake formation |
| YM100 | Boulder EfOM, YM100 permeate | 0.9139 | 0.9880 | 0.9925 | cake formation |
| YM100 | Boulder EfOM | 0.7776 | 0.9594 | 0.9996 | cake formation |
| YM100 | Boulder EfOM, YM100 permeate | 0.9279 | 0.9866 | 0.9898 | --- |
| YM100 | Ruhleben EfOM | 0.7152 | 0.9729 | 0.9966 | cake formation |
| YM100 | Filtrate PP 1 | 0.5792 | 0.9487 | 0.9988 | cake formation |
| YM100 | TPI-A | 0.8918 | 0.9631 | 0.9900 | cake formation |
| YM100 | HPO-A | 0.9045 | 0.9701 | 0.9860 | --- |
| YM100 | SMP-30, GSWP permeate | 0.9651 | 0.9933 | 0.9965 | cake formation and/or pore constriction |
| YM100 | SMP-10, GSWP permeate | 0.9345 | 0.9868 | 0.9995 | cake formation |
| MX500 | Boulder EfOM | 0.7853 | 0.9679 | 0.9997 | cake formation |
| MX500 | Boulder EfOM | 0.7518 | 0.9568 | 0.9997 | cake formation |
| MX500 | Boulder EfOM, MX500 permeate | 0.9882 | 0.9981 | 0.9911 | cake formation and/or pore constriction |
| MX500 | TPI-A | 0.9036 | 0.9719 | 0.9953 | cake formation |
| MX500 | TPI-A + calcium | 0.9473 | 0.9878 | 0.9989 | cake formation |
| MX500 | HPO-A | 0.9504 | 0.9929 | 0.9996 | cake formation and/or pore constriction |

**Table 7.9** (continued)

| Mem-brane | Feed water | Pore blockage ($R^2$-value) | Pore constriction ($R^2$-value) | Cake formation ($R^2$-value) | Predominant fouling mechanism |
|---|---|---|---|---|---|
| GSWP | Boulder EfOM | 0.7266 | 0.9820 | 0.9513 | --- |
| GSWP | Ruhleben EfOM | 0.7149 | 0.9864 | 0.9536 | --- |
| GSWP | Colloids > 6-8 kD (unfiltered) | 0.4354 | 0.9605 | 0.9968 | cake formation |
| GSWP | Colloids > 6-8 kD (pre-filtered) | 0.8845 | 0.9775 | 0.9991 | cake formation |
| GSWP | TPI-A | 0.9497 | 0.9764 | 0.9852 | --- |
| GSWP | HPO-A | 0.9169 | 0.8736 | 0.7863 | --- |
| GSWP | HPO-A | 0.9896 | 0.9998 | 0.9868 | pore constriction |
| GSWP | SMP-30 | 0.9741 | 0.9772 | 0.9891 | --- |
| GSWP | SMP-10 | 0.2069 | 0.8023 | 0.9228 | --- |
| GSWP | SMP-10 | 0.3938 | 0.8061 | 0.9455 | --- |
| GVHP | Boulder EfOM | 0.2595 | 0.9179 | 0.9389 | --- |
| GVHP | Ruhleben EfOM | 0.7160 | 0.9795 | 0.9806 | --- |
| GVHP | Ruhleben EfOM | 0.7035 | 0.9776 | 0.9814 | --- |
| GVHP | Colloids > 12-14 kD (unfiltered) | 0.5555 | 0.9503 | 0.9982 | cake formation |
| GVHP | Colloids > 12-14 kD (pre-filtered) | 0.9550 | 0.9879 | 0.9993 | cake formation |
| GVHP | TPI-A | 0.9597 | 0.9943 | 0.9977 | cake formation and/or pore constriction |
| GVHP | TPI-A | 0.9664 | 0.9917 | 0.9924 | cake formation and/or pore constriction |
| GVHP | HPO-A | 0.8933 | 0.9717 | 0.9923 | cake formation |
| VVLP | Ruhleben EfOM | 0.8170 | 0.9882 | 0.9177 | --- |
| VVLP | Filtrate CAS | 0.7859 | 0.9803 | 0.9843 | --- |
| VVLP | Permeate PP 2 | 0.8412 | 0.9699 | 0.9643 | --- |
| VVLP | Filtrate PP 2 | 0.6182 | 0.9606 | 0.9881 | --- |
| VVLP | Filtrate PP 1 | 0.7427 | 0.9821 | 0.9611 | --- |
| VVLP | Filtrate PP 2 | 0.5597 | 0.9646 | 0.9961 | cake formation |
| VVLP | Filtrate PP 1 | 0.6179 | 0.9574 | 0.9946 | cake formation |
| VVLP | Ruhleben EfOM | 0.5744 | 0.9446 | 0.9961 | cake formation |

An examination of the $R^2$ values for each of the five membranes separately reveals clear differences in the underlying fouling mechanisms for microfiltration and ultrafiltration membranes. The formation of a filtration cake predominates the fouling of the loose hydrophilic UF membrane (YM100) and the tight hydrophilic MF membrane (MX500). This is in accordance with the results of Roorda [2004] who finds cake formation as being the predominant fouling mechanism in pilot scale dead-end ultrafiltration of WWTP effluents. Likewise, Lee [2003] concludes from atomic force microscopy of fouled membranes that surface coverage by natural organic matter predominates in UF membranes. Because of its small pore size of 0.05 μm, the MX500 membrane lies on the border of microfiltration and ultrafiltration membranes. However, in comparison to the other MF membranes used in this research with pore sizes of 0.1-0.22 μm, the MX500 membrane appears to behave more like a loose UF than like the loose MF membranes.

The only exceptions from the predominance of a cake formation with the YM100 and MX500 membranes are the filtration of the hydrophobic acids isolate (HPO-A) and of the permeate samples of the sequential filtration tests. In the case of the UF membrane (YM100) no clear predominance of one of the three classical fouling mechanisms is found for these samples while the flux data of the MF membrane (MX500) can be described by the cake formation model as well as the pore constriction model. Since the permeate samples have already been filtered through an identical membrane (see Section 4.2), it is not surprising that cake formation is not the best or only explanation of the observed fouling behaviour.

While the cake formation model explains the fouling behaviour of loose UF membranes (YM100 and MX500), the mechanisms that predominate loose MF membrane fouling are more complex. In the case of the loose hydrophobic MF membrane (GVHP, pore size 0.22 μm), the fouling behaviour by wastewater treatment plant effluents can not be explained by either of the three classical fouling models, i.e. pore blockage, pore constriction or cake formation. In contrast, the filtration of the isolates with this membrane (GVHP) appears to be dominated by fouling due to cake formation and/or pore constriction.

Neither of the two hydrophilic MF membranes (GSWP and VVLP) shows clear predominance of one of the classical fouling mechanisms. As with the hydrophobic GVHP membrane, the filtration of the colloid isolates with the GSWP membrane (pore size 0.22 μm) represents an exception, exhibiting cake formation as the predominant fouling mechanism. This suggests that the isolated colloids are extremely large macromolecules or are easily forming larger aggregates. Although the last three stirred cell tests with the VVLP membrane (filtrate PP 2, filtrate PP 1, and Ruhleben EfOM) exhibit $R^2$ values of > 0.99 in the linearized plot for the cake formation, the curves of the experimental data points do not look very linear (Figures C.1 – C.3 in Appendix C), especially at the beginning of the membrane filtration tests. Instead the fouling appears to be a combination of several fouling mechanisms.

Ho and Zydney [2000] developed a combined pore blockage – cake formation model to explain the fouling behaviour of a protein (Bovine Serum Albumin) during dead-end microfiltration (see Section 2.1.5). The three parameters of concern are the rate of pore blockage ($\alpha$), the resistance of a single protein aggregate ($R_{pr0}$), and the increase of the hydraulic resistance of the protein deposit over time ($f\cdot R'$). Kilduff et al. [2002] have applied this combined model to natural organic matter nanofiltration, adding a term that accounts for the back-transport due to cross-flow operation. In the case of NOM fouling or EfOM fouling for that matter, $R_{pr0}$ corresponds to the initial resistance of the fouling layer while $f\cdot R'$ represents the change in hydraulic resistance over time due to the fouling layer growth.

A fit of the experimental data of the three loose MF membranes (GVHP, GSWP, and VVLP) with $\alpha$, $R_{pr0}$, and $f\cdot R'$ as fitting parameters gives conflicting results. At first glance, the resulting fit appears to represent the flux decline curve very well (Figure 7.4). However, the corresponding residuals plot, i.e. the difference of the experimental data from the fitted curve (in the direction of the y-axis), exhibits some kind of pattern (Figure 7.5). Ideally, the residuals should randomly scatter around the x-axis at y = 0. Thus, a combination of pore blockage with cake filtration is probably not the best explanation for the observed flux decline. Additionally, the analysis of the three parameters reveals that the fit results in negative values for $R_{pr0}$. This would mean that the foulants actually enhance the permeability instead of reducing it which obviously makes no sense.

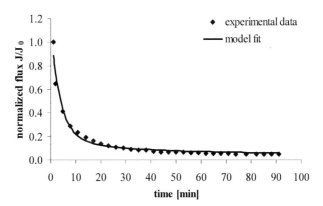

**Figure 7.4**     Fouling behaviour of Ruhleben EfOM with a hydrophilic MF membrane (GSWP): experimental data and fit calculated according to the combined pore blockage – cake formation model of Ho and Zydney [2000]

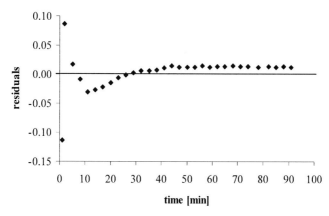

**Figure 7.5**      Residuals plot for model fit shown in Figure 7.4

In summary, it can be said that the fouling of loose UF and tight MF membranes is described by the formation of a filtration cake. In contrast, the fouling of loose MF membranes is more complex and cannot be explained by one of the classical fouling mechanisms alone. However, a model combining pore blockage and cake formation does not explain the flux data, either. More likely is a combination of pore constriction and cake or gel formation as underlying fouling mechanism. This would mean that an adsorption of the foulants occurs in the pores and on the membrane surface. Further in-depth studies of the adsorption kinetics of colloids and polysaccharides on MF membrane surfaces/pores are necessary to validate this hypothesis. Additionally, stirred cell experiments should be automated in order to record more data points at the beginning of the filtration when pore constriction is assumed to dominate the fouling behaviour. This would then allow for the development of a new model combining pore constriction followed by cake or gel formation as predominating fouling mechanisms.

# Chapter 8    Discussion and conclusions

This research investigated the fouling potential of municipal wastewater in low-pressure membrane filtration. The aim was to identify the principal foulants and underlying fouling mechanisms. The fouling behaviour of various feed waters (effluent isolates, bulk effluents, MBR supernatant and permeate) on hydrophobic and hydrophilic microfiltration (MF) and ultrafiltration (UF) membranes was assessed with bench-scale stirred cell experiments. Additionally, two membrane bio-reactor (MBR) pilot plants operating at a municipal wastewater treatment plant (WWTP) were monitored over the course of one year in order to detect possible correlations between the feed water characteristics, operational parameters, and membrane fouling.

## 8.1   *Principal foulants of municipal wastewater in low-pressure membrane filtration*

The results of this research show that the principal foulants of municipal wastewater in low-pressure membrane filtration are hydrophilic substances eluting in the so-called PS peak during size exclusion chromatography (SEC). Compounds eluting in this peak are organic colloids, polysaccharides and large proteins with colloidal character. The substances eluting in the PS peak are likely to be larger than the separation range of the size exclusion column, i.e. > 150,000-200,000 D ($\approx$ 15-20 nm), because they elute close to the void volume of the column. Although the PS peak is completely retained in the monitored MBR pilot plants, one third of the substances eluting in the PS peak are able to pass a similar MF membrane in stirred cell experiments. Thus, these substances are probably smaller than 100-200 nm which is the pore size of the MF membrane (0.1-0.2 µm). Additionally, sequential filtration tests with wastewater effluents suggest that the size of effluent organic matter (EfOM) foulants eluting in the PS peak are in the range of 10-100 nm.

The International Union of Pure and Applied Chemistry (IUPAC) defined in 1971 an object with, in at least one dimension, a size of 1 nm to 1 µm as a colloid. This means that the widely used distinction between particulate and dissolved matter by filtration through 0.45 µm filters not only ignores the colloids but cuts right through the colloidal size range [Hofman 2004]. With sizes between tens to hundreds of nanometers, the fouling causing compounds eluting in the PS peak must be regarded as organic colloids. Bouhabila et al. [2001] determined from hydraulic resistance calculations that colloids and solutes contribute to approximately 75 % to the membrane fouling of a hollow-fiber MF membrane. Roorda [2004] found that the organic matter fraction passing a 0.2 µm filter and being retained by a 0.1 µm filter, is most severely fouling UF membranes in membrane filtration of WWTP effluent. This is slightly larger than the size determined in this research for the colloidal foulants. Nevertheless, in both cases the organic colloids responsible for low-pressure membrane fouling would fall under the definition of dissolved organic matter based on the classical definition used in water chemistry, i.e. < 0.45 µm. Unfortunately, this leads to some confusion about the true character of these compounds.

According to the IUPAC definition of a colloid, humic substances can be seen as organic colloids, too [Buffle and Leppard 1995a; Hofman 2004]. However, size exclusion chromatography revealed that humic substances do not contribute significantly to the membrane fouling nor do hydrophobic organic compounds. Besides organic compounds stemming from drinking water (NOM), soluble microbial products are of relevance for the fouling behaviour of municipal wastewater. This research showed that only approximately 10 % of the effluent organic matter (EfOM) present in wastewater treatment plant effluents cause the fouling of low-pressure membranes. This is in accordance with the findings for NOM fouling of low-pressure membranes, where 5-15 % of the dissolved organic carbon (DOC) are responsible for most of the fouling [Howe and Clark 2002a]. In contrast, the organic carbon content of the PS peak in the supernatant of the MBR pilot plants amounts, on average, to 22 % of the total organic carbon (TOC) content (7 - 45 % for pilot plant 1 and 8 - 34 % for pilot plant 2).

The high fouling potential of the substances eluting in the PS peak during size exclusion chromatography is supported by stirred cell experiments with different fractions of a WWTP effluent. From the three isolated fractions – colloids, hydrophobic acids (HPO-A), and transphilic acids (TPI-A) – the colloids exhibited their main peak in the PS region of the SEC chromatogram and induced the highest flux decline during low-pressure membrane filtration.

## 8.2    Parameters influencing the fouling and the foulants' concentration

The performance of membrane operations is a function of various parameters such as:
-    membrane material and module construction,
-    characteristics of the feed water, and
-    hydrodynamic and operational conditions.

The membrane permeability depends primarily on the interactions between the membrane and the feed water and can be influenced to some extent by the operational and hydrodynamic conditions. Filtration tests with the HPO-A isolate revealed the importance of the membrane material for the fouling. Especially if hydrophobic-hydrophobic interactions between foulants and membrane are expected, a more hydrophilic membrane material should be chosen. Hydrophilic membrane materials exhibit generally a higher pure water permeability which significantly affects membrane filtration as shown by principal component analysis (PCA) of the experimental stirred cell data.

More influential than the membrane properties are the feed water characterisitics on the organic fouling of membranes, namely the characteristics of the organic matter [Her et al. 2004]. The inorganic content is of less importance in low-pressure membranes as scaling is not expected since the divalent ions (salts) pass the membranes unimpeded. For example, a higher flux decline has been observed with the supernatant (filtrate samples) of the MBR pilot plants than with the supernatant (filtrate sample) from the conventional activated sludge system, although the TOC and the organic carbon concentration of the PS peak were similar. This supports the importance of the character of the PS peak, i.e. the composition of this peak (more polysaccharide- or protein-like substances, size of organic colloids).

In the two MBR pilot plants studied, the organic carbon concentration of the PS peak changed over the course of the year resulting in a corresponding change of the total organic carbon with higher values during winter time and lower values in summer. In contrast, the organic carbon concentrations of the humic substances and organic acids peaks remained stable throughout the year at around 11 mg/L. A correlation between the fouling rate of the MBR membrane modules and the polysaccharides concentration as glucose equivalents could be detected for a sludge retention time (SRT) of eight days whereas no clear correlation is found for a sludge retention time of 15 days. Chang and Lee [1998] report an increase in membrane fouling with decreasing SRT.

However, in this research, a correlation between the PS peak (and thus the membrane fouling) and the sludge retention time was not detected. The concentration change in the PS peak appeared to follow the temperature evolution, instead. Especially, the temperature increase in spring corresponds well with the decrease of the PS peak from April 2003 until June 2003 while the SRT is kept constant at eight days during this period. This is supported by principal component analysis which shows that besides the feed water characteristics (TOC, PS content), the temperature accounts for most of the variation in the data. Sludge retention time, on the other hand, does not account for any significant variation.

The low water temperatures during winter ($\sim 15°C$) could also be seen as stress situation for the microorganisms of the activated sludge. It is known from the literature that stress situations for the biomass can enhance the production of extracellular polymeric substances (EPS) by the microorganisms. For example, the addition of toxic substances resulted in an increase of the EPS concentration [Aquino and Stuckey 2004]. Similarly, the shear stress caused by pumps resulting in floc breakage stimulates the production of EPS [Chang et al. 2001]. Thus, the higher PS peak content can be seen as a result of a higher EPS production by the biomass due to lower water temperatures. The EPS is a complex mixture of all kinds of substances such as polysaccharides, proteins, humic substances, and lipids [Jorand et al. 1998], with a significant peak in the PS peak region during size exclusion chromatography and additional peaks in the regions typical for humic substances and organic acids. Further stress situations are, for example, oxygen deprivation or a sudden decrease in sludge retention time. Both situations occurred during the MBR pilot plant operation and resulted in an increase of the amount of substances eluting in the PS peak and a corresponding increase in membrane fouling. This supports the results of Chang and Lee [1998] who observed a higher membrane fouling for higher EPS concentrations in the activated sludge and suggested to use the EPS content as index for membrane fouling in membrane-coupled activated sludge systems.

## 8.3   Fouling mechanisms

Besides the flux decline caused by the fouling due to organic colloids (see above), the reversibility of the fouling layer is of great importance in the efficient operation of membrane plants. The reversibility depends, to a large extent, on the underlying fouling mechanism. Generally, a cake layer is more readily removed by backwash while a gel layer or adsorptive fouling often requires more thorough chemical cleaning. In order to obtain information on the underlying fouling mechanisms, mathematical models can be applied to the flux data gathered during stirred cell experiments. The application of the mathematical model described by Hermia [1982] to the flux data obtained from stirred cell experiments revealed cake formation as being the predominant fouling mechanism for hydrophilic ultrafiltration membranes regardless of the type of feed water (bulk effluent, effluent isolates, MBR supernatant). This is supported by pilot-scale UF membrane filtration of WWTP effluents which exhibited also cake formation as predominant fouling mechanism [Roorda 2004].

In contrast, the fouling of microfiltration membranes is more complex. Ho and Zydney [2000] obtained good results by describing protein fouling of MF membranes with a combined pore blockage and cake formation model. However, neither of the classical fouling models (cake formation, pore blockage, pore constriction [Hermia 1982]) nor the combined pore blockage – cake formation model of Ho and Zydney [2000] could explain the flux data of the stirred cell experiments carried out with MF membranes in this research. Nevertheless, the shape of the curves suggests a transition from one fouling mechanism to another. It is assumed that a combination of pore constriction and cake or gel formation is most likely to best describe the fouling mechanism for microfiltration membranes. This is supported by the observation of Ognier et al. [2002] that fouling due to adsorption (e.g. in the membrane pores) is a rapid phenomenon and occurs at the beginning of the operation.

## 8.4   Conclusions

This research showed that the principal foulants in low-pressure membrane filtration of municipal wastewater are organic colloids. These consist of polysaccharides and large proteins which are part of the extracellular polymeric substances (EPS) produced in the biological wastewater treatment process by the microorganisms of the activated sludge. The amount of EPS released by the biomass depends on the amount of stress inflicted on the microorganisms, e.g. shear stress caused by pumping, low temperature, dissolved oxygen concentration, etc. Hence, it is crucial to maintain constant operating conditions, avoiding any disturbances of the biological treatment, in order to minimize EPS production and thus, membrane fouling.

The mathematical analysis of the stirred cell experiments carried out in this research revealed cake formation as predominant fouling mechanism of UF membranes, while a combination of pore constriction and cake or gel formation is suggested for MF membranes. A comparison of the organic rejection in the MBR pilot plants and in stirred cell tests shows that the organic colloids (PS peak) are completely retained by the pilot plant membrane modules whereas one third of the foulants is able to pass a similar membrane in the bench-scale tests. This supports the hypothesis of pore constriction as an initial fouling mechanism, since it induces a shift of the pore size to smaller sizes (a common phenomenon in MF membrane operations according to Rautenbach [1997]).

## 8.5   Research recommendations

The coincidence of a lower sludge retention time (SRT of 8 days) during winter and a higher sludge retention time (SRT of 15 days) in summer convoluted the data in the case of the MBR pilot plants with regards to the influence of sludge retention time. Experiments with a synthetic wastewater MBR system are recommended for further investigations of the temperature effect and SRT influence on EPS production. The synthetic wastewater MBR system should be designed in a way to allow for the variation of one operating parameter at a time, e.g. either sludge retention time or temperature. It would also be interesting to assess the impact of the biological phosphorous removal in the anaerobic reactor on the production and degradation of EPS in membrane bio-reactors. In this case, a cooperation with microbiologists is suggested to follow the cycle of phosphorous uptake and release by the microorganisms and a possible correlation with the polysaccharides/EPS production and degradation. Additionally, the monitoring of the EPS content in the effluent of a large-scale municipal wastewater treatment plant over the course of at least one year (including all four seasons) could be beneficial. As large-scale plants are usually operated under more stable conditions, seasonal variations are more easily observed.

The assessment of the underlying fouling mechanisms proved difficult due to variations in the data and lack of datapoints primarily at the beginning of the membrane filtration tests. Both can be accounted to the manual data acquisition. Hence, a stirred cell testing system with automated data acquisition is recommended when fouling mechanisms are investigated. This ensures a better database (more data with smaller variations) as a basis for fouling modelling.

Finally, the membrane performance is directly linked to the effectiveness of the applied cleaning regimes. The common use of hypochlorite as chemical cleaning agent does not seem to be ideal as it can cause enhanced membrane aging and leads to the formation of undesirable AOX (adsorbable organic halogens). Cleaning regimes which are better adapted to the specific sorption/desorption characteristics of organic colloids (polysaccharides/ proteins) are needed. This requires more research on the character of the organic colloids and their adsorption kinetics. In order to assess the nature and size of the organic colloids larger size exclusion chromatography (SEC) columns are recommended, e.g. the HW-65S or HW-75S. At the same time more knowledge must be acquired on the inorganic colloids peak in the UV absorption chromatograms in order to answer the question whether the first peak of the OC and UV chromatograms correlate. This could be achieved with known substances of colloidal character and/or large proteins (the standard protein bovine serum albumin is, with $\sim 67,000$ D, an order of magnitude to small).

# Chapter 9    Appendix

## 9.1    Appendix A

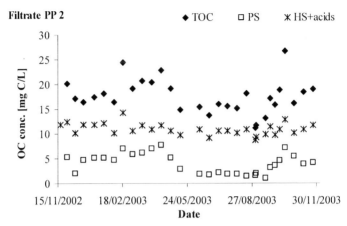

**Figure A.1**    Evolution of total organic carbon (TOC), polysaccharides and organic colloids (PS), and humic substances and organic acids (HS + acids) over a 12 month period for the filtrate of pilot plant 2

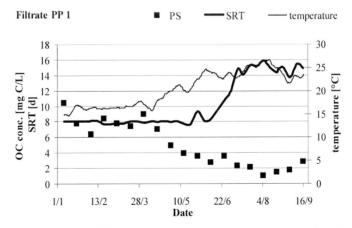

**Figure A.2**    Evolution of PS peak (as organic carbon concentration), sludge retention time (SRT in days), and wastewater temperature (as 7-day running average) from 8.1.-16.9.2003 for pilot plant 1

*9.2*     *Appendix B*

**Table B.1**     PCA database for stirred cell experiments

| membrane | code | feed water | pore size μm | PWP L/hm² | pressure bar | DOC mg/L | UVA$_{254}$ 1/cm | SUVA L/mgm | del. DOC mg C/m² | flux$_{250}$ J/J$_0$ | flux$_{500}$ J/J$_0$ |
|---|---|---|---|---|---|---|---|---|---|---|---|
| YM100 | EfOM | Boulder effluent | 0.01 | 436 | 1.0 | 5.2 | 0.123 | 2.37 | 200 | 0.67 | 0.56 |
| YM100 | EfOM | Boulder effluent | 0.01 | 452 | 1.0 | 5.0 | 0.118 | 2.35 | 272 | 0.76 | 0.54 |
| YM100 | EfOM | UF filt. Boulder eff. | 0.01 | 636 | 1.0 | 4.6 | 0.115 | 2.49 | 2600 | 0.96 | 0.88 |
| YM100 | EfOM | Boulder effluent | 0.01 | 526 | 1.0 | 4.9 | 0.114 | 2.32 | 292 | 0.70 | 0.52 |
| YM100 | EfOM | UF filt. Boulder eff. | 0.01 | 706 | 1.0 | 4.7 | 0.112 | 2.37 | 2400 | 0.96 | 0.92 |
| YM100 | EfOM | Ruhleben effluent | 0.01 | 617 | 1.0 | 15.7 | 0.330 | 2.10 | 295 | 0.79 | 0.59 |
| YM100 | MBRfeed | filtrate PP 1 | 0.01 | 587 | 1.0 | 17.5 | 0.465 | 2.65 | 150 | 0.42 | 0.23 |
| YM100 | TPI-A | TPI-A | 0.01 | 607 | 1.0 | 4.1 | 0.063 | 1.55 | 420 | 0.85 | 0.71 |
| YM100 | HPO-A | HPO-A | 0.01 | 607 | 1.0 | 4.9 | 0.106 | 2.16 | 1400 | 0.99 | 0.90 |
| YM100 | SMP | MF filtered SMP-30 | 0.01 | 743 | 1.0 | 2.1 | 0.136 | 6.48 | 200 | 0.70 | 0.55 |
| YM100 | SMP | MF filtered SMP-10 | 0.01 | 631 | 1.0 | 2.8 | 0.023 | 0.83 | 435 | 0.82 | 0.69 |
| MX500 | EfOM | Boulder effluent | 0.05 | 1007 | 1.0 | 7.6 | 0.121 | 1.60 | 163 | 0.58 | 0.33 |
| MX500 | EfOM | Boulder effluent | 0.05 | 836 | 1.0 | 7.4 | 0.119 | 1.61 | 134 | 0.62 | 0.35 |
| MX500 | EfOM | MF filt. Boulder eff. | 0.05 | 791 | 1.0 | 6.3 | 0.110 | 1.74 | 625 | 0.94 | 0.83 |
| MX500 | TPI-A | TPI-A | 0.05 | 874 | 1.0 | 4.3 | 0.067 | 1.56 | 520 | 0.91 | 0.75 |
| MX500 | TPI-A | TPI-A + Calcium | 0.05 | 880 | 1.0 | 4.5 | 0.074 | 1.65 | 742 | 0.89 | 0.79 |
| MX500 | HPO-A | HPO-A | 0.05 | 838 | 1.0 | 5.0 | 0.109 | 2.18 | 1259 | 1.04 | 0.87 |
| GSWP | EfOM | Boulder effluent | 0.22 | 1276 | 0.3 | 9.1 | 0.136 | 1.50 | 276 | 0.80 | 0.58 |
| GSWP | EfOM | Ruhleben effluent | 0.22 | 1877 | 0.3 | 8.7 | 0.221 | 2.54 | 330 | 0.95 | 0.53 |
| GSWP | COLL | colloids > 6-8 kD | 0.22 | 1747 | 0.3 | 2.2 | 0.035 | 1.59 | 90 | 0.10 | 0.05 |
| GSWP | COLL | colloids > 6-8 kD (filt) | 0.22 | 2096 | 0.3 | 1.6 | 0.009 | 0.57 | 127 | 0.58 | 0.39 |

**Table B.2** PCA database for stirred cell experiments (continued)

| membrane | code | feed water | pore size µm | PWP L/hm² | pressure bar | DOC mg/L | UVA$_{254}$ 1/cm | SUVA L/mgm | del. DOC mg C/m² | flux$_{250}$ J/J$_0$ | flux$_{500}$ J/J$_0$ |
|---|---|---|---|---|---|---|---|---|---|---|---|
| GSWP | TPI-A | TPI-A | 0.22 | 1833 | 0.3 | 5.5 | 0.066 | 1.20 | 2923 | 0.98 | 0.93 |
| GSWP | HPO-A | HPO-A | 0.22 | 1611 | 0.3 | 4.6 | 0.104 | 2.27 | 1850 | 1.07 | 0.97 |
| GSWP | HPO-A | HPO-A | 0.22 | 2134 | 0.3 | 5.0 | 0.103 | 2.07 | 2450 | 1.00 | 0.97 |
| GSWP | SMP | SMP-30 | 0.22 | 1268 | 0.3 | 3.0 | 0.03 | 1.02 | 85 | 0.31 | 0.14 |
| GSWP | SMP | SMP-10 | 0.22 | 1316 | 0.3 | 5.5 | 0.065 | 1.18 | 135 | 0.45 | 0.12 |
| GVHP | EfOM | Boulder effluent | 0.22 | 211 | 1.0 | 9.9 | 0.136 | 1.38 | 20 | 0.20 | 0.19 |
| GVHP | EfOM | Ruhleben effluent | 0.22 | 1661 | 0.6 | 12.7 | 0.536 | 4.22 | 800 | 1.00 | 0.89 |
| GVHP | EfOM | Ruhleben effluent | 0.22 | 1877 | 0.6 | 12.3 | 0.381 | 3.09 | 557 | 1.00 | 0.83 |
| GVHP | COLL | colloids > 12-14 kD | 0.22 | 1552 | 0.6 | 1.9 | 0.034 | 1.76 | 50 | 0.13 | 0.08 |
| GVHP | TPI-A | TPI-A | 0.22 | 741 | 0.6 | 4.9 | 0.078 | 1.59 | 1495 | 1.00 | 0.97 |
| GVHP | TPI-A | TPI-A | 0.22 | 1812 | 0.6 | 4.8 | 0.068 | 1.41 | 2120 | 1.00 | 1.00 |
| GVHP | HPO-A | HPO-A | 0.22 | 1171 | 0.6 | 5.1 | 0.115 | 2.25 | 875 | 0.98 | 0.90 |
| VVLP | EfOM | Ruhleben effluent | 0.1 | 561 | 0.3 | 15.7 | 0.330 | 2.10 | 771 | 1.00 | 0.89 |
| VVLP | CAS | activated sludge | 0.1 | 559 | 0.3 | 18.3 | 0.365 | 1.99 | 529 | 0.90 | 0.80 |
| VVLP | MBRperm | permeate PP 2 | 0.1 | 602 | 0.3 | 13.0 | 0.312 | 2.39 | 3400 | 1.00 | 1.00 |
| VVLP | MBRfeed | filtrate PP 2 | 0.1 | 576 | 0.3 | 16.8 | 0.398 | 2.37 | 320 | 0.84 | 0.35 |
| VVLP | MBRfeed | filtrate PP 1 | 0.1 | 570 | 0.3 | 17.8 | 0.397 | 2.23 | 350 | 0.83 | 0.62 |
| VVLP | MBRfeed | filtrate PP 2 | 0.1 | 591 | 0.3 | 16.3 | 0.400 | 2.45 | 300 | 0.90 | 0.49 |
| VVLP | MBRfeed | filtrate PP 1 | 0.1 | 639 | 0.3 | 14.6 | 0.462 | 3.17 | 300 | 0.84 | 0.47 |
| VVLP | EfOM | Ruhleben effluent | 0.1 | 428 | 0.3 | 9.6 | 0.239 | 2.49 | 137 | 0.45 | 0.19 |

**Table B.3** PCA database for MBR pilot plants: data for pilot plant 1

| Date | TOC mg/L | PS peak mg/L | HS + organic acids peak mg/L | UVA$_{254}$ inorganics peak 1/m | turbidity NTU | SRT days | temperature °C |
|---|---|---|---|---|---|---|---|
| 08/01/2003 | 23.1 | 10.4 | 12.6 | 8.86 | 9.4 | 8.1 | 14.8 |
| 22/01/2003 | 22.8 | 7.8 | 12.0 | 0.55 | 7.2 | 8.0 | 16.9 |
| 06/02/2003 | 19.1 | 6.4 | 10.7 | 4.87 | 5.8 | 8.1 | 16.3 |
| 19/02/2003 | 22.8 | 8.4 | 12.5 | 8.16 | 7.7 | 7.6 | 16.2 |
| 05/03/2003 | 21.2 | 7.8 | 11.3 | 9.41 | 9.0 | 7.8 | 16.4 |
| 19/03/2003 | 22.5 | 7.4 | 11.8 | 5.38 | 7.1 | 8.0 | 16.6 |
| 02/04/2003 | 23.6 | 9.0 | 12.0 | 6.80 | 8.0 | 7.8 | 17.1 |
| 16/04/2003 | 22.0 | 7.0 | 12.0 | 3.85 | 3.1 | 7.9 | 17.9 |
| 30/04/2003 | 22.0 | 4.9 | 11.1 | 1.59 | 4.0 | 8.0 | 19.9 |
| 14/05/2003 | 16.3 | 3.9 | 10.4 | 2.35 | 3.7 | 7.7 | 20.1 |
| 28/05/2003 | 17.3 | 3.6 | 11.3 | 2.69 | 3.8 | 9.3 | 22.5 |
| 11/06/2003 | 17.0 | 2.7 | 11.4 | 3.13 | 5.3 | 8.3 | 24.1 |
| 25/06/2003 | 15.5 | 3.6 | 9.6 | 3.26 | 4.7 | 11.0 | 22.8 |
| 09/07/2003 | 15.5 | 2.3 | 10.2 | 1.10 | 2.7 | 14.8 | 22.9 |
| 22/07/2003 | 15.9 | 2.1 | 11.1 | 1.34 | 5.0 | 15.3 | 25.6 |
| 05/08/2003 | 13.8 | 1.0 | 9.9 | 1.69 | 4.8 | 15.9 | 26.3 |
| 19/08/2003 | 15.5 | 1.5 | 11.7 | 1.34 | 4.4 | 14.3 | 24.9 |
| 02/09/2003 | 11.8 | 1.7 | 9.5 | 2.77 | 7.0 | 13.7 | 21.6 |
| 16/09/2003 | 15.8 | 2.8 | 10.9 | 3.38 | 7.6 | 14.9 | 23.3 |
| 23/09/2003 | 15.4 | 1.8 | 10.7 | 1.07 | 1.5 | 15.5 | 23.4 |
| 30/09/2003 | 11.7 | 1.1 | 8.4 | 1.19 | 2.0 | 15.0 | 21.8 |
| 07/10/2003 | 13.4 | 1.4 | 8.8 | 2.08 | 2.2 | 14.6 | 19.9 |
| 15/10/2003 | 15.5 | 1.3 | 10.1 | 2.64 | 2.9 | 15.4 | 19.4 |
| 28/10/2003 | 13.9 | 1.6 | 9.5 | 1.68 | 4.2 | 15.0 | 15.4 |
| 11/11/2003 | 19.6 | 3.8 | 11.3 | 10.25 | 13.0 | 15.2 | 16.3 |
| 25/11/2003 | 17.5 | 3.0 | 11.7 | 6.57 | 13.0 | 15.6 | 18.9 |

**Table B.4**  PCA database for MBR pilot plants: data for pilot plant 2

| Date | TOC mg/L | PS peak mg/L | HS + organic acids peak mg/L | UVA$_{254}$ inorganics peak 1/m | turbidity NTU | SRT days | temperature °C |
|---|---|---|---|---|---|---|---|
| 08/01/2003 | 17.4 | 5.2 | 11.8 | 6.05 | 5.8 | 8.2 | 15.2 |
| 22/01/2003 | 18.1 | 5.1 | 12.1 | 0.43 | 7.7 | 7.9 | 16.9 |
| 06/02/2003 | 16.5 | 4.7 | 10.2 | 3.15 | 4.2 | 7.9 | 16.3 |
| 19/02/2003 | 24.5 | 7.0 | 14.3 | 9.06 | 11.6 | 7.7 | 17.3 |
| 05/03/2003 | 19.2 | 5.8 | 10.6 | 6.50 | 6.5 | 8.2 | 16.4 |
| 19/03/2003 | 20.7 | 6.1 | 11.7 | 7.96 | 9.6 | 9.4 | 16.0 |
| 02/04/2003 | 20.5 | 7.0 | 10.8 | 6.02 | 7.6 | 9.6 | 16.8 |
| 16/04/2003 | 22.8 | 7.7 | 11.7 | 4.53 | 3.9 | 9.6 | 17.6 |
| 30/04/2003 | 19.1 | 5.2 | 10.6 | 4.70 | 5.6 | 8.7 | 19.6 |
| 14/05/2003 | 14.8 | 2.8 | 9.7 | 3.69 | 5.4 | 8.5 | 19.8 |
| 11/06/2003 | 15.5 | 1.8 | 10.8 | 3.89 | 6.0 | 8.6 | 23.5 |
| 25/06/2003 | 13.7 | 1.7 | 9.1 | 2.47 | 3.6 | 11.4 | 22.4 |
| 09/07/2003 | 16.0 | 2.1 | 10.6 | 1.84 | 4.3 | 14.9 | 22.6 |
| 22/07/2003 | 15.6 | 1.9 | 10.6 | 1.49 | 6.7 | 14.8 | 25.1 |
| 05/08/2003 | 15.1 | 1.9 | 10.1 | 1.50 | 4.4 | 14.9 | 26.3 |
| 19/08/2003 | 18.2 | 1.5 | 10.9 | 1.42 | 3.3 | 14.9 | 24.9 |
| 02/09/2003 | 11.7 | 2.0 | 9.3 | 4.30 | 8.1 | 15.0 | 22.3 |
| 16/09/2003 | 13.2 | 1.0 | 9.8 | 2.27 | 4.9 | 15.1 | 23.3 |
| 23/09/2003 | 17.2 | 3.1 | 11.4 | 2.74 | 4.5 | 15.3 | 22.9 |
| 30/09/2003 | 15.8 | 3.6 | 9.7 | 2.83 | 5.0 | 14.8 | 21.2 |
| 07/10/2003 | 18.9 | 4.6 | 10.8 | 5.52 | 7.1 | 15.8 | 18.9 |
| 15/10/2003 | 26.7 | 7.1 | 12.9 | 19.06 | 12.3 | 14.6 | 18.5 |
| 28/10/2003 | 16.1 | 5.4 | 10.2 | 3.29 | 9.9 | 15.3 | 15.8 |
| 11/11/2003 | 18.5 | 3.9 | 10.9 | 3.14 | 6.4 | 15.5 | 13.9 |
| 25/11/2003 | 19.0 | 4.1 | 11.7 | 4.32 | 24.6 | 14.4 | 18.2 |

## 9.3      *Appendix C*

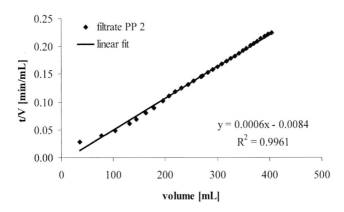

**Figure C.1**    Flux data of the stirred cell experiment using filtrate PP 2 with an MF membrane (VVLP) plotted on characteristic coordinates for cake formation

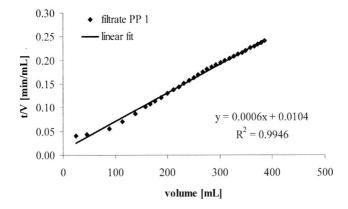

**Figure C.2**    Flux data of the stirred cell experiment using filtrate PP 1 with an MF membrane (VVLP) plotted on characteristic coordinates for cake formation

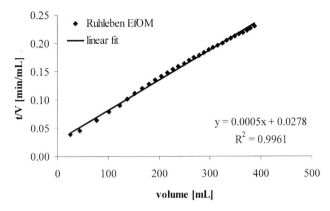

**Figure C.3**   Flux data of the stirred cell experiment using Ruhleben EfOM with an MF
membrane (VVLP) plotted on characteristic coordinates for cake formation

# Chapter 10    References

Aiken, G. (1988): A critical evaluation of the use of macroporous resins for the isolation of aquatic humic substances, in: Humic substances and their role in the environment, eds. F. Frimmel and R. Christman, John Wiley & Sons, pp. 15-28.

Aiken, G. (1985): Isolation and concentration techniques for aquatic humic substances, In: Humic substances in Soil, Sediment, and Water, Aiken, G.; McKnight, D.; Wershaw, R.; and McCarthy, P. (editors), Wiley & Sons, New York, pp. 363-385.

Aiken, G. and Leenheer, J. (1993): Isolation and chemical characterisation of dissolved and colloidal organic matter, Chemistry and Ecology, Vol. 8, pp. 135-151.

Aiken, G.; McKnight, D.; Thorn, K.; and Thurman, E. (1992): Isolation of hydrophilic organic acids from water using nonionic macroporous resins, Org. Geochem., Vol. 18, No. 4, pp. 567-573.

Anselme, C. and Jacobs, E. (1996): Ultrafiltration, Chapter 10, In: Water Treatment Membrane Processes, J. Mallevialle, P. Odendaal, and M. Wiesner (eds.), McGraw-Hill, New York.

Aptel, P. and Buckley, C. (1996): Categories of membrane operations, Chapter 2, In: Water Treatment Membrane Processes, J. Mallevialle, P. Odendaal, and M. Wiesner (eds.), McGraw-Hill, New York.

Aquino, S. and Stuckey, D. (2004): Soluble microbial products formation in anaerobic chemostats in the presence of toxic compounds, Wat. Res., Vol. 38, pp. 255-266.

Barker, D. and Stuckey, D. (1999): A review of soluble microbial products (SMP) in wastewater treatment systems, Wat. Res., Vol. 33, No. 14, pp. 3063-3082.

Bouhabila, E.; Ben Aïm, R.; and Buisson, H. (2001): Fouling characterisation in membrane bio-reactors, Separation/Purification Technology, Vol. 22-23, pp. 123-132.

Buffle, J. and Leppard, G. (1995a): Characterisation of aquatic colloids and macromolecules. 1. Structure and behavior of colloidal material, Environ. Sci. Technol., Vol. 29, No. 9, pp. 2169-2175.

Buffle, J. and Leppard, G. (1995b): Characterisation of aquatic colloids and macromolecules. 2. Key role of physical structures on analytical results, Environ. Sci. Technol., Vol. 29, No. 9, pp. 2176-2184.

Chang, I.-S.; Kim, J.-S.; and Lee, C.-H. (2001): The effects of EPS on membrane fouling in an MBR process, Proceedings MBR 3, 16th May 2001, Cranfield University, England, pp. 19-28.

Chang, I.-S. and Lee, C.-H. (1998): Membrane filtration characteristics in membrane-coupled activated sludge system – the effect of physiological states of activated sludge on membrane fouling, Desalination, Vol. 120, No. 3, pp. 221-233.

Cho, J. (1998): Natural organic matter (NOM) rejection by, and flux decline of, nanofiltration (NF) and ultrafiltration (UF) membranes, Ph.D. thesis, University of Colorado at Boulder, Colorado, USA, cited in Jarusutthirak 2002.

City of Boulder (2002): Annual report 2001, Wastewater Treatment Plant Boulder, CO, USA.

Debroux, J.-F. (1998): The physical-chemical and oxidant-reactive properties of effluent organic matter (EfOM) intended for potable reuse, Ph.D. thesis, University of Colorado at Boulder, USA.

DeCarolis, J.; Hong, S.; Taylor, J.; Naser, S.; Alt, S.; Wilf, M. (2001): Effect of operating conditions on flux decline and removal efficiency of solid, organic and microbial contaminants during ultrafiltration of tertiary wastewater for water reuse, Proceedings AWWA Membrane Conference, San Antonio, Texas, USA.

Defrance, L.; Jaffrin, M.; Gupta, B.; Paullier, P.; and Geaugey, V. (2000): Contribution of various constituents of activated sludge to membrane bio-reactor fouling, Bioresource Technology, Vol. 73, pp. 105-112.

Drewes, J. and Fox, P. (1999): Behavior and characterisation of residual organic compounds in wastewater used for indirect potable reuse, Water Sci. Technol., Vol. 40, No. 4-5, 391-398.

Drewes, J.; Amy, G.; Fox, P.; and Westerhoff, P. (1999): Drinking water TOC impacts on reclaimed water and consequences for regulations of water reuse systems, Proceedings AWWA Annual Conference, Chicago, Illinois, USA.

Dubois, M.; Gilles, K.; Hamilton, J.; Rebers, P.; and Smith, F. (1956): Colorimetric method for determination of sugars and related substances, Analytical Chemistry, Vol. 28, No. 3, pp. 350-356.

EAWAG (2000): www.eawag.ch/research/ing/vt/weiterfuehrende_projekte/joss/proj_aj _memb_d.html, Internet presentation, last updated 21.3.2000.

Eberle, S.; Knobel, K.-P.; and von Hodenberg, S. (1979): Untersuchungen über die Bestimmung des hochmolekularen Anteils des DOC einer Wasserprobe durch Diafiltration, Vom Wasser, Vol. 53, pp. 53-67.

Evenblij, H. and van der Graaf, J. (2003): Occurrence of EPS in activated sludge from a membrane bio-reactor treating municipal wastewater, Proceedings of the conference on Nano and micro particles in water and wastewater treatment, 22.-24.9.2003, Zurich, Switzerland, pp. 329-337.

Everest, W.; Patel, M.; and Alexander, K. (2002): Selecting microfiltration equipment for a water purification mega project, Proceedings of the 5th conference on Membranes in drinking and industrial water production, 22.-26.9.2002, Mülheim an der Ruhr, Germany, supplement of oral and poster presentations pp. 1-7.

Flemming, H.-C.; Schaule, G.; Griebe, T.; Schmitt, J.; and Tamachkiarowa, A. (1997): Biofouling – the Achilles heel of membrane processes, Desalination, Vol. 113, No. 2-3, pp. 215-225.

Frolund, B.; Palmgren, R.; Keiding, K. and Nielsen, P. (1996): Extraction of extracellular polymers from activated sludge using a cation exchange resin, Wat. Res., Vol. 30, pp. 1749-1758.

Gnirss, R. (2004): personal communications.

Gnirss, R.; Lesjean, B.; and Buisson, H. (2003b): Biologische Phosphorentfernung mit einer nachgeschalteten Denitrifikation im Membranbelebungsverfahren, Tagungsband zur 5. Aachener Tagung Siedlungswasserwirtschaft und Verfahrenstechnik, A17, 30.9.-1.10. 2003, Aachen, Germany.

Gnirss, R.; Lesjean, B.; Buisson, H.; Adam, Chr.; Kraume, M. (2003a): Enhanced biological phosphorus removal (EBPR) in membrane bio-reactors (MBR), Proceedings of the AWWA Membrane Technology Conference, 2.-5. March 2003, Atlanta, USA.

GROM Analytik + HPLC GmbH (2003): Catalogue, bulk materials, gel permeation, p. 141, www.grom.de/products/framprod1.htm.

Habarou, H.; Makdissy, G.; Croué, J.-P.; Amy, G.; Buisson, H.; and Machinal, C. (2001): Toward an understanding of NOM fouling of UF membranes, Proceedings AWWA Membrane Conference, San Antonio, Texas, USA.

Her, N.; Amy, G.; Foss, D.; Cho, J.; Yoon, Y.; and Kosenka, P. (2002): Optimization of method for detecting and characterizing NOM by HPLC-size exclusion chromatography with UV and on-line DOC detection, Environ. Sci. Technol., Vol. 36, pp. 1069-1076.

Her, N.; Amy, G.; Park, H.-R.; Song, M. (2004): Characterizing algogenic organic matter (AOM) and evaluating associated NF membrane fouling, Wat. Res., Vol. 38, pp. 1427-1438.

Hermia, J. (1982): Constant pressure blocking filtration laws – application to power-law non-newtonian fluids, Transactions of the Institution of Chemical Engineers, Vol. 60, pp. 183-187.

Ho, C.-C. and Zydney, L. (2000): A combined pore blockage and cake filtration model for protein fouling during microfiltration, J. of Colloid and Interface Science, Vol. 232, pp. 389-399.

Hofmann, T. (2004): Die Welt der vernachlässigten Dimensionen: Kolloide, Chem. Unserer Zeit, Vol. 38, pp. 24-35.

Howe, K. and Clark, M. (2002a): Coagulation pretreatment for membrane filtration, AWWARF, Denver, USA.

Howe, K. and Clark, M. (2002b): Fouling of microfiltration and ultrafiltration membranes by natural waters, Environ. Sci. Technol., Vol. 36, No. 16, pp. 3571-3576.

Huber, S. and Frimmel, F. (1991): Flow injection analysis of organic and inorganic carbon in the low ppb range, Analytical Chemistry, Vol. 63, pp. 2122-2130.

Huber, S. and Frimmel, F. (1996): Gelchromatographie mit Kohlenstoffdetektion (LC-OCD): Ein rasches und aussagekräftiges Verfahren zur Charakterisierung hydrophiler organischer Wasserinhaltsstoffe, Vom Wasser, Vol. 86, pp. 277-290.

Huber, S. and Gluschke, M. (1998): Chromatographic characterisation of TOC in process water treatment. Ultrapure Water, March 1998, pp. 48-52.

Ishiguro, K.; Imai, K.; and Sawada, S. (1994): Effects of biological treatment conditions on permeate flux of UF membrane in a membrane/activated-sludge wastewater treatment system, Desalination, Vol. 98, pp. 119-126.

Jacangelo, J. and Buckley, C. (1996): Microfiltration, Chapter 11, In: Water Treatment Membrane Processes, J. Mallevialle, P. Odendaal, and M. Wiesner (eds.), McGraw-Hill, New York.

Jarusutthirak, C. (2002): Fouling and flux decline of reverse osmosis (RO), nanofiltration (NF), and ultrafiltration (UF) membranes associated with effluent organic matter (EfOM) during wastewater reclamation/reuse, Ph.D. thesis, Department of Civil, Environmental, and Architectural Engineering, University of Colorado at Boulder, Colorado, USA.

Jarusutthirak, C.; Amy, G.; Drewes, J.; and Fox, P. (2002): Nanofiltration and ultrafiltration membrane filtration of wastewater effluent organic matter (EfOM): Rejection and fouling, Proceedings of IWA congress, 3rd World Water Congress, Melbourne, Australia.

Jolliffe, I. (1986): Principal component analysis, Springer series in statistics, Springer-Verlag, New York.

Jorand, F.; Boué-Bigne, F.; Block, J.; and Urbain, V. (1998): Hydrophobic/hydrophilic properties of activated sludge exopolymeric substances, Wat. Sci. Tech., Vol. 37, No. 4-5, pp. 307-315.

Kilduff, J.; Mattaraj, S.; Sensibaugh, J.; Pieracci, J.; Yuan, Y.; and Belfort, G. (2002): Modeling flux decline during nanofiltration of NOM with poly(arylsulfone) membranes modified using UV-assisted graft polymerization, Environ. Eng. Sci., Vol. 19, No. 6, pp. 477-495.

Laabs, C.; Amy, G.; and Jekel, M. (2003): Organic colloids and their influence on low-pressure membrane filtration, Proceedings of the conference on Nano and micro particles in water and wastewater treatment, 22.-24.9.2003, Zurich, Switzerland, pp. 321-327.

Laabs, C.; Amy, G.; Jekel, M.; and Buisson, H. (2002): Fouling of low-pressure (MF and UF) membranes by wastewater effluent organic matter (EfOM): Characterisation of EfOM foulants in relation to membrane properties, Proceedings of the 5th conference on Membranes in drinking and industrial water production, 22.-26.9.2002, Mülheim an der Ruhr, Germany, pp. 693-700.

Laspidou, C. and Rittmann, B. (2002): A unified theory for extracellular polymeric substances, soluble microbial products, and active and inert biomass, Wat. Res., Vol. 36, No. 11, pp. 2711-2720.

Lee, N. (2003): Natural organic matter (NOM) fouling of low-pressure (MF and UF) membranes: Identification of foulants, fouling mechanisms, and evaluation of pretreatment, Ph.D. thesis, Department of Civil, Environmental, and Architectural Engineering, University of Colorado at Boulder, Colorado, USA.

Lee, W.; Kang, S.; and Shin, H. (2003): Sludge characteristics and their contribution to microfiltration in submerged membrane bio-reactors, J. Mem. Sci., Vol. 216, pp. 217-227.

Leenheer, J.(2002): personal communication.

Leenheer, J.; Croué, J.-P.; Benjamin, M.; Korshin, G.; Hwang, C.; Bruchet, A.; and Aiken, G. (2000): Comprehensive isolation of natural organic matter from water for spectral characterisations and reactivity testing, In: Natural organic matter and disinfection by-products, Barrett, S.; Krasner, S.; and Amy, G. (editors), ACS Symposium Series 761, Washington, DC, Chapter 5, pp. 68-83.

Lowry, O.; Rosebrough, N.; Farr, A.; and Randall, R. (1951): Protein measurement with the folin phenol reagent, Journal of Biological Chemistry, Vol. 193, No. 1, pp. 265-275.

Madaeni, S.; Fane, A.; and Wiley D. (1999): Factors influencing critical flux in membrane filtration of activated sludge, J. Chem. Technol. Biotechnol., Vol. 74, pp. 539-543.

Manem, J. and Sanderson, R. (1996): Membrane bio-reactors, Chapter 17, In: Water Treatment Membrane Processes, J. Mallevialle, P. Odendaal, and M. Wiesner (eds.), McGraw-Hill, New York.

Menkveld, H.; Slange, J.; and Durieux, F. (2003): Immersed ultrafiltration for tertiary treatment and reuse of WWTP effluent, Tagungsband zur 5. Aachener Tagung Siedlungswasserwirtschaft und Verfahrenstechnik, W16, 30.9.-1.10.2003, Aachen, Germany.

Mulder, M. (1991): Basic principles of membrane technology, Chapter VII, Kluwer Academic Publishers, Dordrecht, The Netherlands.

Ognier, S.; Wisniewski, C.; Grasmick, A. (2002): Influence of macromolecule adsorption during filtration of a membrane bio-reactor mixed liquor suspension, J. Mem. Sci., Vol. 209, pp. 27-37.

Parameshwaran, K.; Fane, A.; Cho, B.; and Kim, K. (2001): Analysis of microfiltration performance with constant flux processing of secondary effluent, Wat. Res., Vol. 35, No. 18, pp. 4349-4358.

Rauch, T. (2003): personal communications.

Rautenbach, R. (1997): Membranverfahren, Springer Verlag, Berlin, Germany.

Reichenbach, C.; Jekel, M.; and Amy, G. (2001): Rejection of wastewater effluent organic matter by microfiltration and ultrafiltration membranes for potable reuse, Proceedings AWWA Water Quality Technology Conference, Nashville, Tennessee, USA.

Reith, C. and Birkenhead, B. (1998): Membranes enabling the affordable and cost effective reuse of wastewater as an alternative water resource, Desalination, Vol. 117, pp. 203-210.

Ridgway, H. and Flemming, H.-C. (1996): Membrane biofouling, Chapter 6, In: Water Treatment Membrane Processes, J. Mallevialle, P. Odendaal, and M. Wiesner (eds.), McGraw-Hill, New York.

Roorda, J. (2004): Filtration characteristics in dead-end ultrafiltration of wwtp-effluent, Ph.D. thesis, TU Delft, The Netherlands.

Roorda, J. and van der Graaf, J. (2003): Filtration characteristics in dead-end ultrafiltration of WWTP-effluent, Tagungsband zur 5. Aachener Tagung Siedlungswasserwirtschaft und Verfahrenstechnik, A17, 30.9.-1.10. 2003, Aachen, Germany.

Rosenberger, S. (2003): Charakterisierung von belebtem Schlamm in Membran-belebungsreaktoren zur Abwasserreinigung, Fortschritt-Berichte VDI, Reihe 3, Nr. 769, VDI Verlag, Düsseldorf, Germany.

Rosenberger, S.; Lesjean, B.; Laabs, C.; Jekel, M.; Gnirss, R.; and Schrotter, J.-C. (2004): Impact of colloidal and soluble organic material on membrane performance in membrane bio-reactors for municipal wastewater treatment, submitted for publication.

Sato, S. and Ishii, Y. (1991): Effects of activated sludge properties on water flux of ultrafiltration membrane used for human excrement treatment, Wat. Sci. Technol., Vol. 23, pp. 1601-1608.

Schlegel, H. (1992): Allgemeine Mikrobiologie, 7. überarbeitete Auflage, Georg Thieme Verlag, Stuttgart, Germany.

Schumacher, J. (2002): unpublished results.

Schwedt, G. (1996): Taschenatlas der Umweltchemie, Georg Thieme Verlag, Stuttgart, New York.

Shin, H.-S. and Kang, S.-T. (2003): Characteristics and fates of soluble microbial products in ceramic membrane bio-reactors at various sludge retention times, Wat. Res., Vol. 37, pp. 121-127.

Shin, H.-S.; Lee, W.-T.; Kang, S.-T.; Park, H.-S.; and Kim, J.-O. (2002): Contribution of solids and soluble materials of sludge to UF behavior under starvation, Proceedings IMSTEC 2002, Melbourne, Australia.

Skoog, D.; Holler, F.; and Nieman, T. (1998): Principles of Instrumental Analysis, Chapter 17, 5th edition, Harcourt Brace College Publishers, Philadelphia, USA.

Späth, R. (1998): Rolle der extrazellulären polymeren Substanzen in Biofilm und belebtem Schlamm bei der Sorption von Schadstoffen, Dissertation am Lehrstuhl für Wassergüte- und Abfallwirtschaft an der Technischen Universität München, Germany.

StatSoft, Inc. (2000): STATISTICA for Windows, Computer Program Manual, Tulsa, OK, USA.

Stephenson, T.; Judd, S.; and Jefferson, B. (2003): Membrane technology for wastewater treatment and reuse – state of the art and future perspectives, Tagungsband zur 5. Aachener Tagung Siedlungswasserwirtschaft und Verfahrenstechnik, Ü4, 30.9.-1.10.2003, Aachen, Germany.

Taylor, J. and Jacobs, E. (1996): Reverse osmosis and nanofiltration, Chapter 9, In: Water Treatment Membrane Processes, J. Mallevialle, P. Odendaal, and M. Wiesner (eds.), McGraw-Hill, New York.

Tchobanoglous, G.; Darby, J.; Bourgeous, K.; McArdle, J.; Genest, P.; and Tylla, M. (1998): Ultrafiltration as an advanced tertiary treatment process for municipal wastewater, Desalination, Vol. 119, pp. 315-322.

Tsuru, T.; Urairi, M.; Nakao, S.-I.; and Kimura, S. (1991): Negative rejection of anions in the loose reverse osmosis separation of mono- and divalent ion mixtures, Desalination, Vol. 81, No. 1-3, pp. 219-227.

Voßenkaul, K. and Rautenbach, R. (1997): Konzepte für Ultrafiltrationsanlagen – Modulbauformen und Betriebsweisen, Tagungsband zur 1. Aachener Tagung Siedlungswasserwirtschaft und Verfahrenstechnik, A5, 30.6.-1.7.1997, Aachen, Germany.

Wiesner, M. and Aptel, P. (1996): Mass transport and permeate flux and fouling in pressure-driven processes, Chapter 4, In: Water Treatment Membrane Processes, J. Mallevialle, P. Odendaal, and M. Wiesner (eds.), McGraw-Hill, New York.

Wijmans, J. and Baker, R. (1995): The solution-diffusion model: a review, J. Membrane Science, Vol. 107, No. 1-2, pp. 1-21.

Yuan, W. and Zydney, A. (2000): Humic acid fouling during ultrafiltration, Environ. Sci. Technol., Vol. 34, No. 23, pp. 5043-5050.

Yuan, W. and Zydney, A. (1999): Effects of solution environment on humic acid fouling during microfiltration, Desalination, Vol. 122, pp. 63-76.